5th International Conference on Manufacturing, Material and Metallurgical Engineering (ICMMME 2020)

AA007557

IOP Conference Series: Materials Science and Engineering
Volume 859

Osaka, Japan
28 – 31 March 2020

ISBN: 978-1-7138-2687-3
ISSN: 1757-8981

Printed from e-media with permission by:

Curran Associates, Inc.
57 Morehouse Lane
Red Hook, NY 12571

Some format issues inherent in the e-media version may also appear in this print version.

This work is licensed under a Creative Commons Attribution 3.0 International Licence.
Licence details: http://creativecommons.org/licenses/by/3.0/.

No changes have been made to the content of these proceedings. There may be changes to pagination and minor adjustments for aesthetics.

Printed with permission by Curran Associates, Inc. (2021)

For permission requests, please contact the Institute of Physics at the address below.

Institute of Physics
Dirac House, Temple Back
Bristol BS1 6BE UK

Phone: 44 1 17 929 7481
Fax: 44 1 17 920 0979

techtracking@iop.org

Additional copies of this publication are available from:

Curran Associates, Inc.
57 Morehouse Lane
Red Hook, NY 12571 USA
Phone: 845-758-0400
Fax: 845-758-2633
Email: curran@proceedings.com
Web: www.proceedings.com

TABLE OF CONTENTS

MATERIAL CHARACTERIZATION AND PHYSICAL PROPERTIES

MATERIAL STRUCTURE DESIGN AND PERFORMANCE ANALYSIS

MATERIALS CHEMISTRY AND MEASUREMENT

ENGINEERING MATERIAL PROCESSING AND PREPARATION TECHNOLOGY

DIGITAL MANUFACTURING TECHNOLOGY AND MECHANICAL SYSTEMS

IOP Conf. Series: Materials Science and Engineering **859** (2020) 011001 doi:10.1088/1757-899X/859/1/011001

PREFACE

It was nevertheless a great pleasure for me as conference chair to welcome all the participants in the 2020 5th International Conference on Manufacturing, Material and Metallurgical Engineering (ICMMME 2020) conducted online, 28 - 31 March 2020. In spite of this major change in conducting the conference, it was gratifying to note the participation of researchers from all over the world. ICMMME is dedicated to new research developments and advances in the fields of Manufacturing, Materials and Metallurgical Engineering.

The 5th International Conference on Manufacturing, Material and Metallurgical Engineering (ICMMME 2020) is an annual conference since 2016 and it was supposed to be held in Osaka, Japan, 28 - 31 March 2020. Unfortunately, due to Coronavirus pandemic, it's hard to predict when the epidemic will be effectively controlled globally since our participants are from different countries and regions, taken the successful experiences from some influential international conferences as well as the suggestions from most of the participants, we decided to hold ICMMME 2020 virtually.

To ensure this virtual conference goes smoothly, ICMMME 2020 was well prepared under the technical support of Chengdu Young Academic Conference Co. Ltd.. One-to-one software testing was conducted before the formal meeting, the video presentations were collected for back up to avoid technical problem such as bad network. These videos were destroyed after conference to protect intellectual property. Technical service was provided also throughout the whole conference. The 4 days' program was highlighted with 3 keynote speeches and 6 parallel sessions including 6 presentations for each. Participants shared their current research remotely and reported present original ideas or results of general significant supported by clear reasoning and compelling evidence, and employ methods, theories and practices relevant to the research.

The proceeding of the conference consists of papers from the universities, research institutes and industry. The papers were selected after a rigorous peer-review by conference committee members and international reviewers. The papers should provide the readers an overview of many recent advances in the fields related to materials especially the composite materials and metallurgical engineering including the modeling, analysis, computation and optimization as well as the experimental methods.

I would like to thank all the authors who have contributed to this volume and also to the members of the organizing committee, reviewers, speakers, chairpersons, sponsors and the conference participants for their support to ICMMME 2020.

Content from this work may be used under the terms of the Creative Commons Attribution 3.0 licence. Any further distribution of this work must maintain attribution to the author(s) and the title of the work, journal citation and DOI.
Published under licence by IOP Publishing Ltd

IOP Conf. Series: Materials Science and Engineering **859** (2020) 011001 doi:10.1088/1757-899X/859/1/011001

ICMMME Conference Chair
Prof. Ramesh K. Agarwal
Washington University in St. Louis, USA

Peer review statement

All papers published in this volume of *IOP Conference Series: Materials Science and Engineering* have been peer reviewed through processes administered by the proceedings Editors. Reviews were conducted by expert referees to the professional and scientific standards expected of a proceedings journal published by IOP Publishing.

Conference Chairs

Prof. Ramesh K. Agarwal, Washington University in St. Louis, USA
Prof. Changduk Kong, Chosun University, South Korea

Program Chairs

Prof. Tsutomu Yoshida, Takushoku University, Japan
Prof. Seong Soo Choi, SunMoon University, South Korea
Prof. S. Lee, Gyeongsang National University, South Korea

Local Arrangement Chairs

Prof. Mikio Ito, Osaka University, Japan
Prof. Takashi Takuma, Osaka Institute of Technology, Japan

Workshop Chair

Prof. Hyunbum Park, Kunsan National University, South Korea

Technical Committee

Prof. Md Mustafizur Rahman, Universiti Malaysia Pahang, Malaysia
Prof. Poi Sim Khiew, Universiti of Nottingham, Malaysia
Prof. Datin Napsiah Ismail, Universiti Putra Malaysia, Malaysia
Prof. Recai KUS, Selcuk University, Turkey
Prof. Michel Bellet, Institute Mines-Telecom. Mines ParisTech, France
Prof. Pavel Lukáč, Univerzita Karlova, Czech Republic
Prof. Abel Ribeiro de Jesus, Federal University of Bahia, Brazil
Prof. Nicoleta GILLICH, "Eftimie Murgu" University of Resita, Romania
Prof. Zdzislaw Adamczyk, Silesian University of Technology, Poland
Prof. Paul Mativenga, The University of Manchester, UK
Prof. Ali Arslan Kaya, Mugla Sitki Kocman University, Turkey
Prof. Farid Abed, American University of Sharjah, United Arab Emirates
Prof. Manyi Chen, Wuhan University of Technology, China
Prof. Christopher R. Bowen, University of Bath, UK
Prof. Xin-Lin Gao, Southern Methodist University, USA
Prof. Khalil Abdelrazek Khalil Abdelmawgoud, University of Sharjah, United Arab Emirates
Prof. Laichang Zhang, Edith Cowan University, Australia
Prof. Norhamidi Muhamad, Universiti Kebangsaan Malaysia, Malaysia
Prof. Megat Mohamad Hamdan bin Megat Ahmad, Universiti Pertahanan Nasional Malaysia
Prof. Jean-Bernard VOGT, The University of Lille, France
Prof. Gilbert-Rainer GILLICH,"EftimieMurgu"University Resita, Romania
Assoc. Prof. MELLAL Mohamed Arezki, M'‘HamedBougaraUniversity(UMBB), Algeria

Content from this work may be used under the terms of the Creative Commons Attribution 3.0 licence. Any further distribution of this work must maintain attribution to the author(s) and the title of the work, journal citation and DOI.
Published under licence by IOP Publishing Ltd

IOP Conf. Series: Materials Science and Engineering **859** (2020) 011002 doi:10.1088/1757-899X/859/1/011002

Assoc. Prof. Rosli Ahmad, Universiti Tun Hussein Onn Malaysia, Malaysia
Assoc. Prof. Thanakom Soontornchainacksaeng, King Mongkut's University of Technology North Bangkok, Thailand
Assoc. Prof. Chou Chau Yuan Fong, Universiti Brunei Darussalam, Brunei
Assist. Prof. Zhibin Lin, North Dakota State University, USA
Assist. Prof. Muhammad Jahan, Miami University, USA
Assoc. Prof. Raj Das, RMIT University, Australia
Assoc. Prof. NAKAMURA Yuzo, Kagoshima University, Japan
Assist. Prof. Nyuk Yoong Voo, Universiti Brunei Darussalam, Brunei Darussalam
Assoc. Prof. MA Xianfeng, Sun Yat-sen University, China
Assoc. Prof. Mohd Rasidi Bin Haji Ibrahim, Universiti Tun Hussein Onn Malaysia, Malaysia
Assoc. Prof. Kojiro Iizuka, Shibaura Institution of Technology, Japan
Assoc. Prof. Rusdi Nur, Politeknik Negeri Ujung Pandang, Indonesia
Assoc. Prof. Cheng-Yang Liu, National Yang-Ming University, Taiwan
Assist. Prof. AN Liang, Hong Kong Polytechnic University
Assist. Prof. Michele Dassisti, Politecnico di Bari, Italy
Dr. Cristian Paul Chioncel,"Eftimie Murgu" University of Resita, Romania
Dr. Jinqiang Ning, Georgia Institute of Technology, Atlanta, USA
Dr. Quanli Han, Xi'an Technological University, China
Dr. Elammaran Jayamani, Swinburne University of Technology Sarawak Campus, Malaysia
Dr. Amar Rezoug, Superieure National School of Technology (ENST), Algeria
Dr. Hoong Thiam Toh, Universiti Teknologi Malaysia, Malaysia

IOP Conf. Series: Materials Science and Engineering **859** (2020) 011003 doi:10.1088/1757-899X/859/1/011003

First Name	Last Name	Institution
Vladimir	Filatov	Bauman Moscow State Technical University, Russia
Ardian Agus	Permana	Prince of Songkla University, Thailand
Svetlana	Pichkurenko	Bauman Moscow State Technical University, Russia
Chih-Ping	Wu	National Cheng Kung University, Taiwan
Grich	Kongphet	Khon Kaen University, Thailand
Isadora	Costa	Khon Kaen University, Thailand
Irina	Garkina	Penza State University of Architecture and Construction,
Junfeng	Wang	China Univesity of Petroleum (East China) , China
Thepwachara	Ruchirabha	Nagaoka University of Technology, Japan
Pattarabordee	Khaigunha	Khon Kaen University, Thailand
Thidarat	Meekoch	Chulalongkorn University, Thailand
Thanakornkan	Limlertchareonwanit	Chulalongkorn University, Thailand
Thanadol	Hinthao	Khon Kaen University, Thailand
Aibao	Chai	University of Nottingham Malaysia
Kunthisa	Jehlaeh	Prince of Songkla University, Thailand
Akihito	Itami	Graduate School of Doshisha University, Japan
Jakapan	Pimolrat	Chulalongkorn University, Thailand
Leandro	Silverio	Aeronautics Institute of Technology (ITA), Brazil
Mohamed M.A.	Ammar	Monash University, Australia
Cheng-Yang	Liu	National Yang-Ming University, Taiwan
Leandro	Silverio	Aeronautics Institute of Technology (ITA), Brazil

Content from this work may be used under the terms of the Creative Commons Attribution 3.0 licence. Any further distribution of this work must maintain attribution to the author(s) and the title of the work, journal citation and DOI.
Published under licence by IOP Publishing Ltd

Email Address
filatov-vladimir@yandex.ru
ardianaguspermana@gmail.com
swetlathok@gmail.com
cpwu@mail.ncku.edu.tw
kogrich@kkumail.com
costa.isadora@hotmail.com
i.a.naum@mail.ru
wjfgmj@gmail.com
s177011@stn.nagaokaut.ac.jp
pattarabordee.lpt@gmail.com
Krengkrai.M@chula.ac.th
Krengkrai.M@chula.ac.th
Thanadol_h@kkumail.com
aibao.chai@nottingham.edu.my
fah.kunthisa@gmail.com
ctwc0520@gmail.com
Krengkrai.M@chula.ac.th
leandro.silverio.pereira@gmail.co
mohamed.ammar@monash.edu
cyliu66@ym.edu.tw
leandro.silverio.pereira@gmail.co

IOP Conf. Series: Materials Science and Engineering **859** (2020) 011004 doi:10.1088/1757-899X/859/1/011004

Peer review statement

All papers published in this volume of *IOP Conference Series: Materials Science and Engineering* have been peer reviewed through processes administered by the proceedings Editors. Reviews were conducted by expert referees to the professional and scientific standards expected of a proceedings journal published by IOP Publishing.

Content from this work may be used under the terms of the Creative Commons Attribution 3.0 licence. Any further distribution of this work must maintain attribution to the author(s) and the title of the work, journal citation and DOI.
Published under licence by IOP Publishing Ltd

IOP Conf. Series: Materials Science and Engineering **859** (2020) 012001 doi:10.1088/1757-899X/859/1/012001

Iodine Plasmonic Crystal as the Visible-Range Spectral Filter

V V Filatov, V S Gorelik, S V Pichkurenko

Bauman Moscow State Technical University, Moscow, Russia

filatov-vladimir@yandex.ru

Abstract. We analyse the optical properties of the new hyperbolic metamaterial based on the one-dimensional mesoporous aluminum oxide (Al_2O_3) photonic crystal film doped by metallic iodine (I_2) nanoparticles. The nanocomposite is a metal-dielectric hybrid and combines the classical photonic crystal features (such as photonic bandgaps, light deceleration, negative refraction etc.) with the metallic ones (low-frequency mirroring, plasmons etc.). To get this, we propose to saturate the pores (microcavities) of the periodic aluminum oxide photonic crystal matrix synthesized by the controlled acid etching at its anonization, by the iodine vapor. This way, in the photonic crystal pores, we form the cold plasma of the iodine free electrons, and make the nanocomposite the plasmonic photonic one. Note, that the iodine saturation level defines the plasma frequency which is directly affects the crystal characteristics, so we get them easy-tunable. For example, we offer the two iodine-saturated photonic crystals with various doping level in the pass-and-reflect scheme to spectral filter the optical radiation. At last, we also wish to note that simplicity of the saturation technology makes the iodine plasmonic crystals attractive for the wide commercial usage in the large-scale optic and photonic applications and devices (such as spectral filters, optical amplifiers, high-performance selective mirrors, photonic sensors etc.).

1. Introduction

Nowadays, there is a worldwide interest to the new photonic materials. The photonic crystals – nanocomposites with the photonic bandgap at the visible spectral region – are the most prominent of them due to their unique features, such as the negative refraction, the light deceleration etc. [1]

The simplest photonic crystal is the 1D Bragg stack of the identical material layers with the regular gaps between [2]. In the case of metal layers, the metal free electrons in layers form the cold plasma, and the photonic crystal becomes the plasmonic one. Therefore, we get a new metamaterial to be investigated.

2. Analysis

To deal with the plasmonic crystal, we use the Bloch-Floquet formalism [3,4]. This way, the electromagnetic waves in the 1D periodic medium describes by the dispersion equation [5]:

$$\cos ka = \cos k_1 a_1 \cdot \cos k_2 a_2 - \frac{1}{2}\left(\frac{n_1}{n_2} + \frac{n_2}{n_1}\right) \cdot \sin k_1 a_1 \cdot \sin k_2 a_2. \quad (1)$$

Here a_1 and a_2 are, respectively, the odd and the even layers' thicknesses, n_1 and n_2 are its' refractive indices, $k_i = \omega \cdot n_i / c$ ($i = 1,2$) are the wavenumbers, ω is the wavefrequency and $c = 3 \cdot 10^8$ *m/s* is the speed of light.

Content from this work may be used under the terms of the Creative Commons Attribution 3.0 licence. Any further distribution of this work must maintain attribution to the author(s) and the title of the work, journal citation and DOI.
Published under licence by IOP Publishing Ltd

In the plasmonic crystal, there are the metallic (i = 1) layers and the dielectric (i = 2) ones in-between.

For the metallic layer, we have the cold electron plasma with the refractive index [6]

$$n_1 = \sqrt{1 - \frac{\omega_p^2}{\omega^2}}, \tag{2}$$

where ωp = qe·(ne/meε0)1/2 is the plasma frequency, qe = 1.6·10-19 C is the elementary charge, me = 9.1·10-31 kg is the electron mass, ε0 = 8.85·10-12 F/m is the electric constant and ne is the electron concentration. And because of metal, ne = n0·v, where n0 is the atom concentration and v is the metal's chemical valence. For example, for the crystalline iodine I2 [7] n0 = N/V = 2/(325·10-30) = 6.2·1027 m-3, v = 2 and, therefore, ωp = 6·1015 rad/s. So, the typical ωp is about 1015 rad/s, and the plasmons are the optical ones.

As for the dielectric layers, their refractive index

$$n_2 = const \tag{3}$$

is the index of refraction of the corresponding material and can varies from n2 = 1 (void) to almost infinity (organic polymers).

Anyway, the composite reflection (at the normal incidence) is [8]

$$R(\omega) = \left|\frac{n(\omega)-1}{n(\omega)+1}\right|^2 = \left|\frac{k(\omega)-\omega/c}{k(\omega)+\omega/c}\right|^2, \tag{4}$$

the group velocity of light in the composite is

$$\mathrm{v}(\omega) = \frac{d\omega}{dk} = \left[\frac{dk(\omega)}{d\omega}\right]^{-1}, \tag{5}$$

and the effective refractive index is

$$n(\omega) = [c \cdot k(\omega)/\omega] \cdot sign\, \mathrm{v}(\omega) \tag{6}$$

with k(ω) is defined by (1).

3. Simulation

The simulation results for the iodine-sapphire plasmonic crystal are presented at the fig. 1-4. The crystal parameters are $a_1 = a_2 = 100\ nm$, $\omega_p = 6\cdot 10^{15}\ rad/s$ (iodine) and n_2 =1.77 (sapphire).

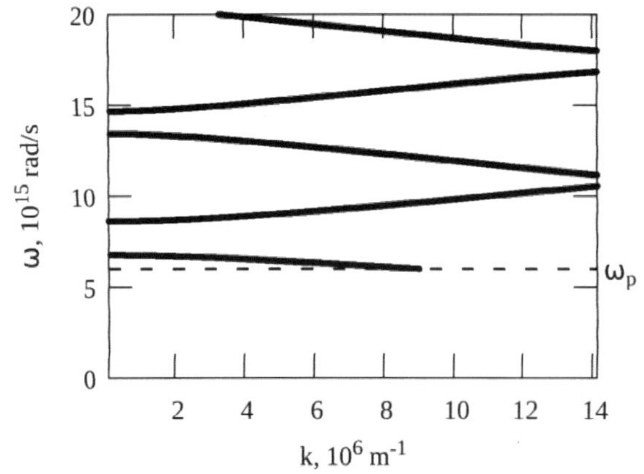

Figure 1. The dispersion law (1) for the electromagnetic waves in the iodine-sapphire plasmonic crystal. Note the low-frequency (zero to ω_p) bandgap due to plasmons.

IOP Conf. Series: Materials Science and Engineering **859** (2020) 012001 doi:10.1088/1757-899X/859/1/012001

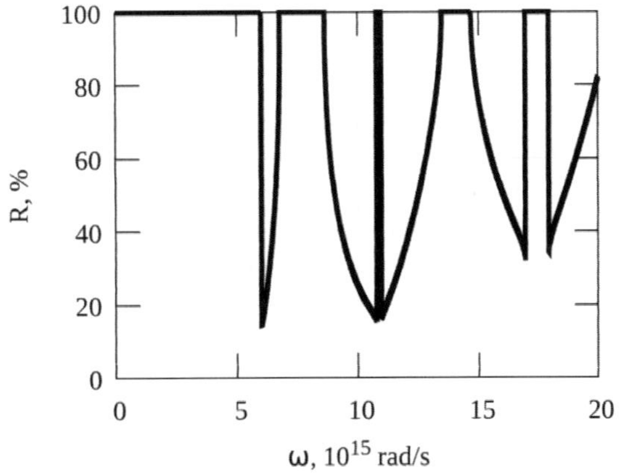

Figure 2. The reflectance (4) of the plasmonic crystal. In the visible range, there are several 100%-reflectance bands due to the photonic bandgaps. The widest low-frequency band is the plasmonic one.

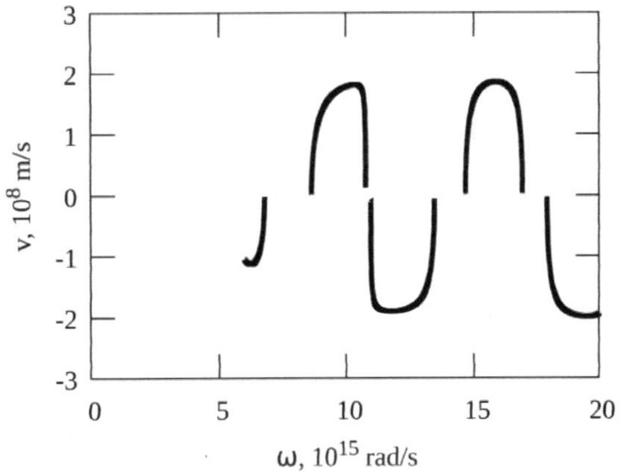

Figure 3. The group velocity (5) of light in the plasmonic crystal. No-solution areas are photonic bandgaps. Note the light deceleration to the zero at the bandgap edges.

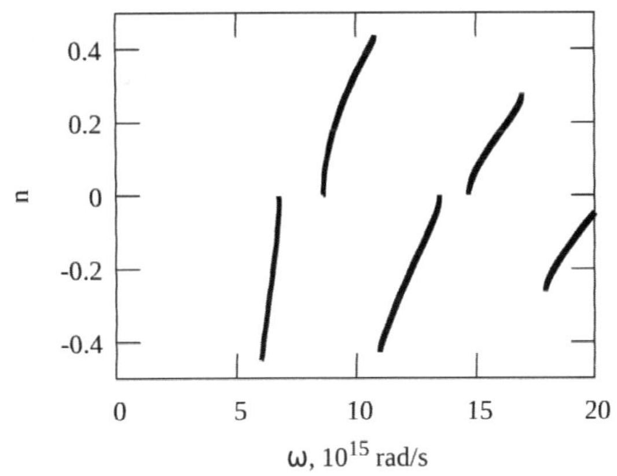

Figure 4. The effective refractive index of the nanocomposite. Due to plasmons $|n| < 1$ (see (2)).

IOP Conf. Series: Materials Science and Engineering **859** (2020) 012001 doi:10.1088/1757-899X/859/1/012001

4. Discussion

The simulation gives (see fig. 1) that in the plasmonic crystal, the acoustic mode cuts itself at the plasma frequency. This forms the additional low frequency (zero to the plasma frequency) photonic bandgap to reflect the light (fig. 2). This way, two plasmonic crystals with different plasma frequencies can act as a spectral filter, fig. 5.

At the fig. 5, one can see, that when the crystal no. 2 are rotated at the small angle (for the convenience, at the fig. 5 the angle is exaggerated), we have the pass-and-reflect scheme to filter the electromagnetic radiation. For example, when the crystal no. 1 plasma frequency is $\omega_p = 2.4 \cdot 10^{15}$ *rad/s* ($\lambda = 800$ *nm*) and the crystal no. 2's one is $\omega_p = 4.7 \cdot 10^{15}$ *rad/s* ($\lambda = 400$ *nm*), we get the visible range filtered at whole.

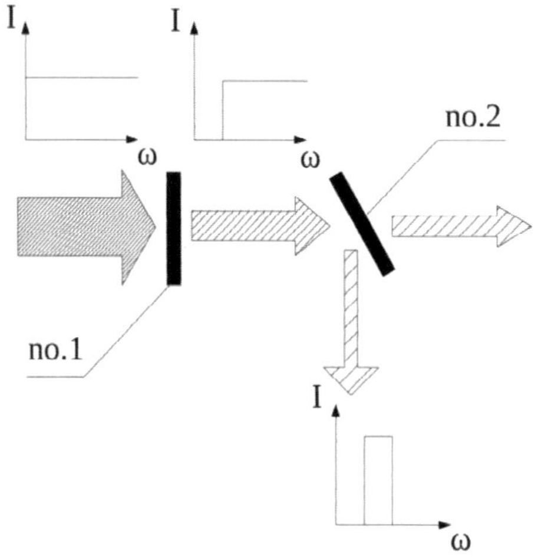

Figure 5. Two plasmonic crystals (at the fig., no. 1 and no. 2) act as a spectral filter.

Note, that the plasmonic crystal's plasma frequency can be tuned by the doping of crystal's metallic layers by the dielectric inclusions. This dilutes the metal concentration and lessen the plasma frequency. To do this, we propose the inverse technique: to saturate the pores of layer-forming dielectric matrix by the iodine vapor (our experiments [9,10] show that solid metals are hard to enter), to create the "metallic" layers. This way, the mesoporous aluminum-based photonic crystal films [11-17] are the best to deal with, because of its porous structure, fig. 6.

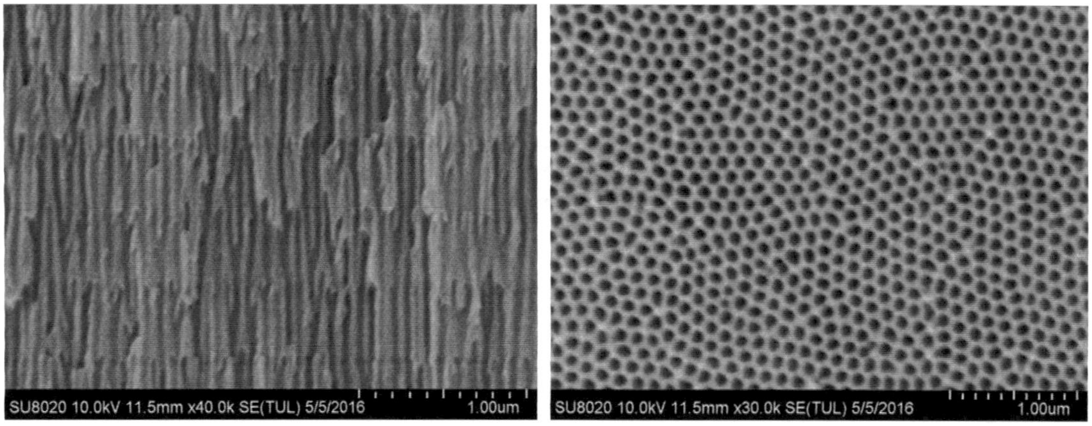

a B

Figure 6. The SEM image of Al_2O_3 photonic crystal film: (*a*) slice with horizontal layer periodicity, (*b*) top view with a lot of pores of 10 *nm* in a diameter and 100 *nm* in a length (from our paper [16]).

5. Conclusion

The results show that the iodine-saturated mesoporous aluminium-based photonic crystal is the new hyperbolic (see (2)) metamaterial, the plasmonic photonic crystal, combines the photonic crystal (specific dispersion, fig. 1*;* photonic bandgaps, fig. 2; slow light, fig. 3*;* negative refraction, fig. 4) and metal (low-frequency mirroring) features. This makes it perspective for the various photonic applications (such as the electromagnetic field amplifiers, e.g. for the enhancement of the Raman scattering [17]; high-efficiency selective mirrors, photonic sensors, etc.) and, besides, for the spectral filtering of an optical radiation. Moreover, the simplicity of technology (vapour saturation of the dielectric matrix) makes it attractive for the wide commercial usage.

Acknowledgements
The work is supported by the Russian Fund for Basic Researches (grant no. 18-02-00181).

References

[1] V S Gorelik (2007) *Quantum Electronics* 37 409.
[2] V P Bykov (1972) *Sov. Phys. JETP* 35 269.
[3] G Floquet (1883) *Annales Scientifique de l'École Normale Supé rieure* 12 47.
[4] F Bloch (1928) *Z. Phys.* 52 555.
[5] A Yariv and P Yeh (1984) *Optical Waves in Crystals: Propagation and Control of Laser Radiation* (Wiley, New York).
[6] F F Chen (1984) *Introduction to Plasma Physics and Controlled Fusion. Volume 1: Plasma Physics (*Springer US).
[7] R M Ibberson, O Moze and C Petrillo (1992) *Molecular Physics: An International Journal at the Interface Between Chemistry and Physics* 76 395.
[8] M Born M and E Wolf (1999) *Principles of Optics: Electromagnetic Theory of Propagation, Interference and Diffraction of Light, 7th ed.* (Cambridge University Press).
[9] V S Gorelik, V V Filatov (2012) *Inorganic Materials* 48 361.
[10] Y P Voinov, V S Gorelik, L I Zlobina, V V Filatov (2009) *Inorganic Materials* 45 1133.
[11] M Ashurov, V Gorelik, K Napolskii and S Klimonsky (2019) *Photonic Sensors* DOI:10.1007/s13320-019-0569-2
[12] V S Gorelik, Dongxue Bi, Guang Tao Fei, Shao Hui Xu and Xu Dong Gao (2019) *Inorganic Materials* 55 355.
[13] V S Gorelik, M M Yashin, Dungxue Bi and Guang Tao Fei (2018) *Optics and Spectroscopy* 124 167.
[14] V S Gorelik, P P Sverbil, V V Filatov, Dongxue Bi, Guang Tao Fei and Shao Hui Xu (2018) *Photonics and Nanostructures – Fundamentals and Applications.* 32 6.
[15] V S Gorelik, M M Yashin and A I Vodchits (2017) *Physics of Wave Phenomena* 25 175.
[16] V S Gorelik, Yu P Voinov, V V Shchavlev, Dongxue Bi, Guo Liang Shang and Guang Tao Fei (2017) *Journal of Surface Investigation: X-ray, Synchrotron and Neutron Techniques* 11 246.
[17] V S Gorelik, A V Pudovkin and V V Filatov (2016) *Journal of Russian Laser Research* 37 604.

IOP Conf. Series: Materials Science and Engineering **859** (2020) 012002 doi:10.1088/1757-899X/859/1/012002

Discharge Energy Efficiency Improvement of P(VDF-HFP) Copolymers Thin Films by Stretching and Electron Beam Irradiation

A A Permana, C Putson

Department of Physics, Faculty of Science, Prince of Songkla University, Hat Yai, Songkhla, 90110, Thailand

chatchai.p@psu.ac.th

Abstract. Toward improving the applications for energy-based technology, dielectric polymers is getting attention due to its relatively high dielectric constant, dielectric breakdown and flexibility, with easily preparation, lightweight and low cost. Dielectric contribution and polarization responses lead to different shape and size of hysteresis loop. This work presents the techniques on reducing domain size for slimmer loop, indicating lower energy loss. As-casted P(VDF-HFP), stretched P(VDF-HFP) and electron irradiated-stretched P(VDF-HFP) thin films were prepared by solution casting method. The irradiation was prepared by emitting electron beam. The dielectric and AC conductivity properties were investigated by LCR meter, while polarization-electric field loops were observed by P-E loop instrument. The results show that stretching and electron beam irradiation significantly increase the dielectric constant of P(VDF-HFP). Their ability on modifying the domain size leads to reduce P-E loop of P(VDF-HFP), followed by reducing energy loss but improving storage energy density and discharge energy efficiency.

1. Introduction

The outstanding materials for current applications in renewable energy-based electronics [1], [2] and power electronics [3] cannot be separated from its high dielectric constant [4] as well as low dielectric loss. One of the applications of dielectric polymers related to energy-based electronics is capacitor [5]. Since invented in 1990, dielectric polymers film capacitor with higher capacitance stability, owning high-voltage characteristics and possessing extremely low losses [6], has been improved due to its advantages over electrolytic and ceramic capacitor. In power and energy-based applications, low dielectric loss and high dielectric constant could lead to a higher energy efficiency and better performance as well.

Beyond its advantages that can be fabricated into thin film and large size device, P(VDF)-based semi-crystalline polymers [7] are the best materials that containing all electroactive characteristics among other ferroelectric, piezoelectric and pyroelectric materials [8]. With addition of hexafluoropropylene (HFP) into the main chain of P(VDF), P(VDF-HFP) copolymers is rich on stability, chemically resistance, non-toxicity and ability on tailoring the size and shape as well as recycling capability [9], [10].

Based on ferroelectricity of P(VDF)-based polymers including P(VDF-HFP) copolymers, a normal ferroelectric dielectric materials polarization response on applied external electric field will exhibit a

Content from this work may be used under the terms of the Creative Commons Attribution 3.0 licence. Any further distribution of this work must maintain attribution to the author(s) and the title of the work, journal citation and DOI.
Published under licence by IOP Publishing Ltd

broad hysteresis loop [11]. That phenomenon leads to a high energy loss and low storage energy density. As the consequence, the discharge energy efficiency is also drop. This work focuses on increasing storage energy density and energy efficiency with lowering energy loss at the same time.

2. Materials and methods

2.1. Materials and method

P(VDF-HFP) Solef 11010/1001 copolymers powder was supplied by Solvay Solexis, Belgium. DMF N,N-dimethylformamide with ≥ 99% purity was supplied by RCI Labscan, Thailand.

First condition, poly(vivylidene fluoride-hexafluoroprophylene) thin films, named as-casted P(VDF-HFP) were kindly prepared by solution casting. The P(VDF-HFP) powder was stirred with DMF solvent in room temperature using magnetic stirrer for 16 hour. The solution then been filled out and casted on the glass plate using blade with the thickness of 30 ± 5 μ m. The samples then been dried for 12 hour at 80 °C in the oven. The second condition, named stretched P(VDF-HFP) was prepared by stretching P(VDF-HFP) films from 120 ± 5 μ m to 30 ± 5 μ m by stretching machine at 70 °C. Lastly, the stretched film then been emitted by 15 kV electron beam using conventional scanning electron microscopy (SEM) HITACHI TM3030Plus from Germany for 5 minutes. This condition named electron beam irradiated-stretched P(VDF-HFP) or EI stretched P(VDF-HFP).

2.2. Characterizations and measurements

LCR meter IM 3533 HIOKI was carried out for measuring dielectric constant and dielectric loss as well as AC conductivity. It was conducted on 1 V of AC voltage with varying the range of frequency from 1 Hz to 100 kHz at room temperature. The Dielectric constant and AC conductivity can be generated by calculating the data of the capacitance (*C*) and conductance (*G*) that gained from the measurements. Meanwhile, the value of dielectric loss can be gained directly from the measurements.

The hysteresis loops of polarization and electric field were measured by P-E loop instrument Trek type 601E. Conducted in temperature range from 21 °C to 140 °C, a 40 MV m^{-1} of external electric field with 10 Hz of frequency were applied to the thin films. Those P-E hysteresis loops were compared between each condition by calculating the area inside the loop (ferroelectric energy loss) and the area outside of the loop (ferroelectric energy density). Furthermore, the discharge energy efficiency can be obtained for each condition. Mathematically, it can be expressed as:

$$\eta = \frac{U_e}{U_e + U_l} \quad (1)$$

where η, U_e, and U_l refers to energy efficiency, energy density and energy loss, respectively [12].

3. Results and discussion

3.1. LCR meter results

The dielectric constant of as-casted P(VDF-HFP), stretched P(VDF-FP) and EI stretched P(VDF-HFP) are presented by figure 1. Figure 1 clearly shows that as-casted P(VDF-HFP) possesses the lowest value of dielectric constant. As reported by other works, as-casted P(VDF-HFP) is dominated by α-phase that is the most stable crystalline phase due to its symmetric structures [13]. On the other hands, the dielectric constant is increase significantly after mechanical stretching in every frequency. There is no doubt that mechanical stretching is one of the most effective ways to increase dielectric constant of the materials. It is due to the ability of mechanical stretching on facilitating the transformation of non-polar α-phase to the most useful β-phase [11]–[14]. Furthermore, mechanical stretching is also responsible to rearrange the crystal grains orientation and increase its uniformity [15]. Those mechanism will leads to the more order β-phase dipole TTTT conformation of P(VDF-HFP).

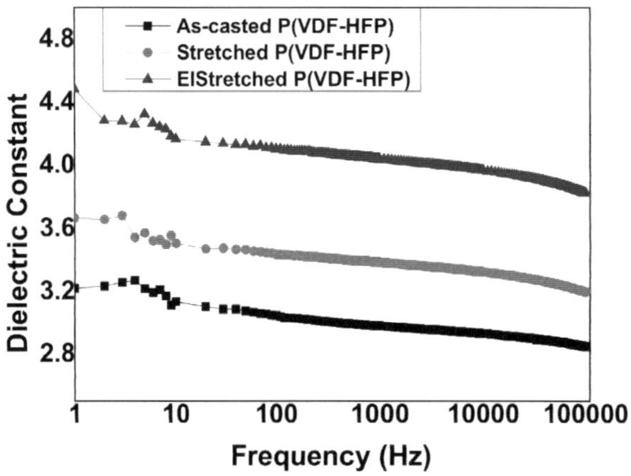

Figure 1. Dielectric constant of as-casted, stretched and EI stretched P(VDF-HFP) as a function of electric field frequency.

The dielectric constant improvement of P(VDF-HFP) caused by electron beam irradiation on stretched P(VDF-HFP) is about two times compared than that of the only stretched thin films in all frequencies. This shows that both mechanical stretching and electron beam irradiation are effective to modify the structure of the thin film. As the results, the dielectric constant are increase significantly compare with as-casted condition. For example, dielectric constant of as-casted, stretched and EI stretched at 1 Hz frequency are 3.22, 3.66 and 4.48 respectively.

Dielectric loss is relatively referred to AC conductivity. Normally, the improvement of dielectric constant will be followed by the increasement of dielectric loss and AC conductivity also. That is a challenge that usually found in every dielectric material.

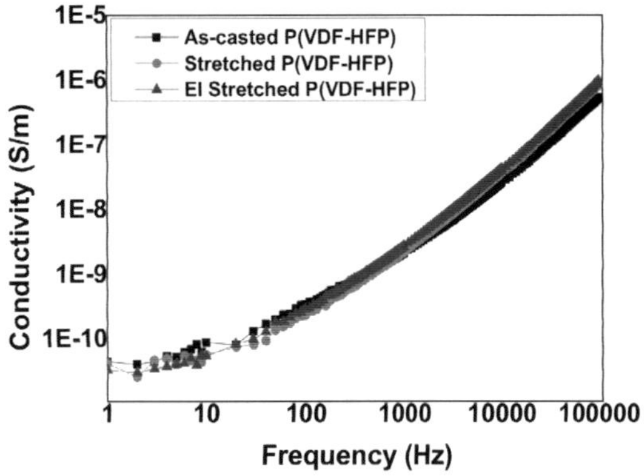

Figure 2. AC conductivity of as-casted, stretched and EI stretched P(VDF-HFP) as a function of electric field frequency.

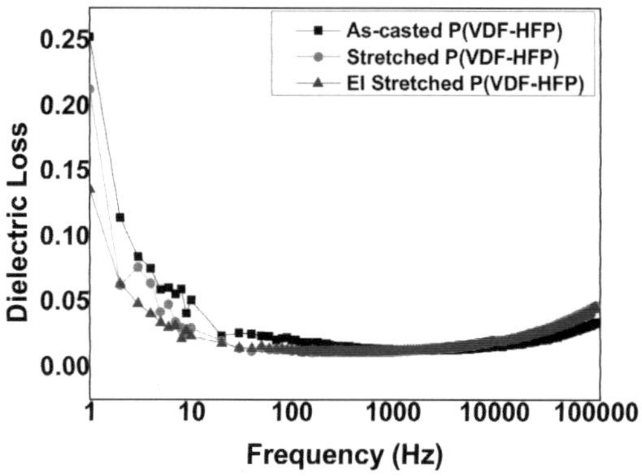

Figure 3. Dielectric loss of as-casted, stretched and EI stretched P(VDF-HFP) as a function of electric field frequency.

Nevertheless, figure 2 and 3 show that AC conductivity and dielectric loss of P(VDF-HFP) are not drastically change after stretching and electron beam irradiation. Somehow, at low frequency the numbers are even gradually drops. Some previous works reported that stretching [11], [15] is responsible to the improvement of dielectric constant due to its ability to convert non-polar α-phase to polar β-phase. At the same time, stretching also effects to destroy the big ferroelectric domain of P(VDF-HFP). Hence, the crystalline size/ferroelectric domain after stretching tends to be smaller as reported by Li and Wang in 2016 [11]. As a final result, mechanical stretching could control the dielectric loss or even decrease its number, while increasing the dielectric constant at the same time.

Similar with the effect of stretching on AC conductivity and dielectric loss, electron beam irradiation is also responsible to stabilize those two parameters. It is well known that electron beam irradiation is becoming an alternative way to boost dielectric constant and keep control or even drop the AC conductivity and dielectric loss. Its ability is associated with its role when cutting the long chain to be shorter [16]. Furthermore, generating smaller crystalline size, electron beam irradiation would be very useful on generating smaller P-E loop that refers to smaller energy loss.

3.2. P-E loop measurements

Figure 4 shows P-E loop at room temperature (28 °C) with the external electric field of 40 MV/m. It shows clearly that stretched and electron beam irradiated stretched P(VDF-HFP) possess higher maximum polarization (P_{max}). P_{max} is increase from 0.56 for as-casted film to 0.64 and 0.65 μ C/cm^2 for stretched and EI stretched films, respectively. At the same time, the remnant polarizations (P_r) of those two conditions are smaller than that of as-casted P(VDF-HFP). P_r is decrease from 0.09 for as-casted film to 0.04 and 0.02 μ C/cm^2 for stretched and EI stretched films, respectively.

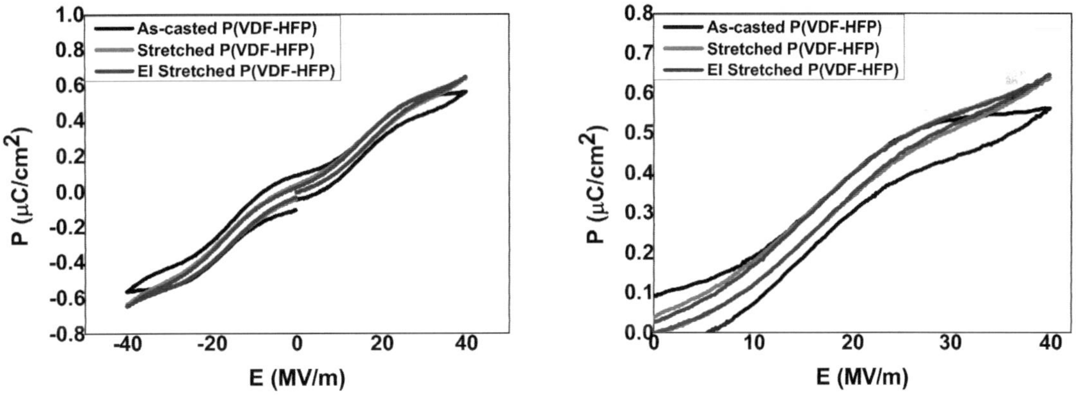

Figure 4. P-E loop of as-casted, stretched and EI stretched P(VDF-HFP).

The bigger loop of as-casted films refers to large size of the ferroelectric domain that consist of big number of dipoles, causing higher remnant polarization [12], [17]. The dropping of P_r that caused by stretching and electron irradiation effects to the polarization responses on external electric field, exhibiting the slimmer loops. In another word, stretching and electron beam irradiation are effectively reduce the area inside the loop of the P(VDF-HFP) thin films, confirming the previous work [15].

3.3. Energy efficiency

Based on P-E loop for each condition, the storage energy density, energy loss as well as energy efficiency value have been obtained. The results were shown by figure 5. It is clear that stretching and electron beam irradiation are effective to boost the storage energy density of P(VDF-HFP) thin films. At the same time, both treatments also drop the energy loss, drastically. The combination of these two parameters effects to the energy efficiency itself, as shown by figure 6. Hence, the discharge energy efficiency tends to exhibit much better value. The results of P-E loop measurements are also confirmed by dielectric constant result. In 2012, Zhu was reported that storage energy density can be improved by improving dielectric constant and electrical breakdown strength [12], [18].

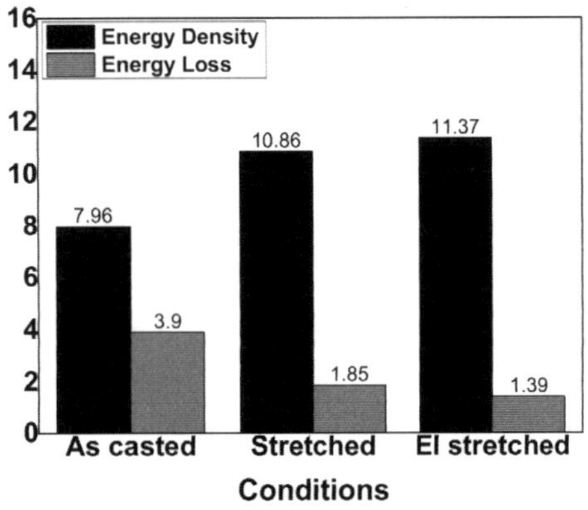

Figure 5. Energy density and loss of as-casted, stretched and EI stretched P(VDF-HFP).

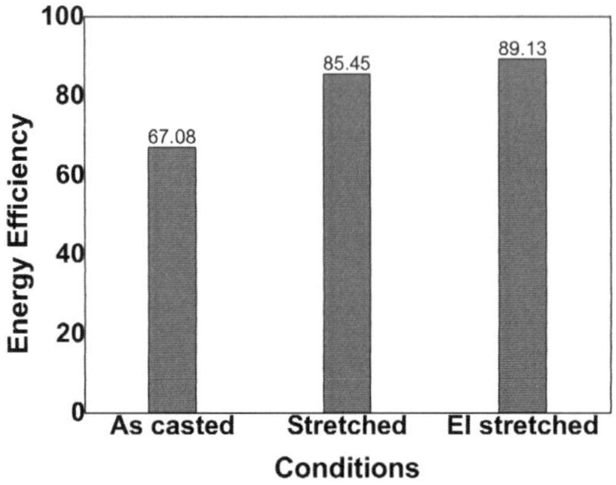

Figure 6. Energy efficiency of as-casted, stretched and EI stretched P(VDF-HFP).

4. Conclusion

Stretching and electron beam irradiation effectively improve the dielectric constant of P(VDF-HFP) copolymers thin films. The improvement was observed at all range of frequency. It corresponds to the ability of both mechanical stretching as well as electron beam irradiation on changing the crystalline phase from nonpolar α-phase to the polar β-phase. It also effect on changing the polarization-electric field loop (P-E loop) of the P(VDF-HFP). The improvement of maximum polarization and drops of remnant polarization cause the hysteresis loop getting slimmer. It represents that energy loss is decrease while storage energy density and discharge energy efficiency are increase at the same time. Hence, these properties are good to be applied on energy-based electronics.

Acknowledgments

This work was supported by Department of Physics, Faculty of Science, Prince of Songkla University (PSU) Thailand, Thailand Center of Excellence in Physics (ThEP-61-PIP-PSU3) and Institute of Biomedical Engineering, Faculty of Medicine, Prince of Songkla University (PSU) Thailand.

References

[1] Satake A, Jike J and Mitani Y 2019 *Int. J. Elec. & Elecn. Eng. & Telcomm.* 8 146
[2] Lim Y S, Chok E T, Chua K H and Liew S 2018 *Int. J. Elec. & Elecn. Eng. & Telcomm.* 7 38
[3] Abdukhakimov A, Bhardwaj S, and Kim D 2019 *Int. J. Elec. & Elecn. Eng. & Telcomm.* 8 233
[4] Jung H 2019 Int. J. Elec. & Elecn. Eng. & Telcomm. 8 334
[5] Qiao Y, Yin X, Zhu T, Li H and Tang C 2018 *Prog. Polym. Sci.* 80 153
[6] Baer E and Zhu L 2017 *Macromolecules.* 50 2239
[7] Liu J, Lu X and Wu C 2013 *Membranes.* 3 389
[8] Martins P, Lopes A C and L-Mendez S 2014 *Prog. Polym. Sci.* 39 683
[9] Roy S, Thakur P, Hoque N, Bagchi B and Das S 2016 *RSC Adv.* 6 21881
[10] Thakur V K and Gupta R K 2016 *Chem. Rev.* 116 4260
[11] Li Q and Wang Q 2016 *Macromol. Chem. Phys.* 217 1228
[12] Zhu Y, Jiang P, Zhang Z and Huang X 2017 *Chinese Chem. Lett.* 28 2027
[13] He F A *et al* 2016 *Compos. Sci. Technol.* 137 138
[14] Tiwari V K *et al* 2013 *Macromolecules.* 46 5595
[15] Tan S, Hu X, Ding S, Zhang Z, Li H and Yang L 2013 *J. Mater. Chem. A.* 1 10353
[16] Zhu Y, Jiang P, Zhang Z and Huang X 2017 *Chinese Chem. Lett.* 28 2027
[17] Cheng Z Y, Olson D, Xu H, Xia F, Hundal J S and Zhang Q M 2002 *Macromolucules.* 35 664
[18] Zhu L and Wang Q 2012 *Macromolecules.* 45 2937

IOP Conf. Series: Materials Science and Engineering **859** (2020) 012003 doi:10.1088/1757-899X/859/1/012003

Dispersion of Light in the 1D Photonic Crystal

S V Pichkurenko

Bauman Moscow State Technical University, Moscow, Russia

swetlathok@gmail.com

Abstract. We propose a novel, Kurosawa-like model to evaluate the 1D (Bragg stack-like) mesoporous aluminium oxide photonic crystal. To do this, we analyze the internal potential of the photonic crystal superlattice and get it describing the set of the medium's polar oscillators. Unlike the atomic oscillators for a common crystal, these ones are the abstract ones. This way, the real photonic crystal can be dealt as an abstract oscillators' ensemble. The result is fully agreed with the thermodynamics, and makes the theory very powerful. To obtain the oscillators parameters, we compare the theory with the secondary emission spectrum of the crystal, and get the natural frequency and the force for each oscillator. This phenomenological approach allow us to calculate photonic crystal's optical characteristics, such as the dispersion law for the light in the nanostructure, the secondary emission spectrum of the composite, the speed of light in the crystal and the effective mass of the speed quanta. We establish the room-temperature Bose-Einstein condensation of polaritons in crystal at the photonic bandgap edge. The results are important to the solid-state detection of paraphotons.

1. Introduction

Nowadays, there is a worldwide interest to the photonic crystals and films, the nanocomposites with photonic bandgaps in the visible and near-IR specrtal range [1]. In the paper, we construct the novel phenomenological model to analyze its optical characteristics.

2. Analysis

To derive the dispersion law for the light in the 1D photonic crystal, let's expand the internal photonic crystal potential $V(x)$ near the equilibrium point $x = 0$:

$$V(x) = V(0) + V'(0)x + \frac{V''(0)}{2}x^2 + \ldots = \frac{V''(0)}{2}x^2 + o(x^2) \quad \equiv \frac{m_0\omega_0^2}{2}x^2 + o(x^2) \quad . \qquad (1)$$

Because of this, we can substitute the real material medium by a set of abstract material oscillators. In the optical spectral range, these are the electronic oscillators [2] with the following dynamics:

$$\ddot{x} + m_0\omega_0^2 x = q_0 E_0 e^{i\omega t}. \qquad (2)$$

Here x is the electron coordinate, m_0 is the mass of the oscillator, q_0 is its charge, ω_0 is its natural frequency and $E_0e^{i\omega t}$ is the external field (in general, ω can differ from ω_0). The solution for (2) is

$$x(t) = \frac{q_0 E_0 e^{i\omega t}}{m_0(\omega_0^2 - \omega^2)}. \qquad (3)$$

Content from this work may be used under the terms of the Creative Commons Attribution 3.0 licence. Any further distribution of this work must maintain attribution to the author(s) and the title of the work, journal citation and DOI.
Published under licence by IOP Publishing Ltd

To get the dispersion equation, let's use the macroscopic parameters. For the material medium with n_0 polar oscillators per unit volume, we have the medium polarization

$$P = n_0 q_0 x \equiv (\varepsilon - 1)\varepsilon_0 E. \tag{4}$$

By substitute (4) to (3), we get

$$\varepsilon(\omega) = 1 + \frac{n_0 q_0^2}{m_0 \varepsilon_0 (\omega_0^2 - \omega^2)} \equiv 1 + \frac{\omega_p^2}{\omega_0^2 - \omega^2} \equiv \frac{\omega_l^2 - \omega^2}{\omega_0^2 - \omega^2}. \tag{5}$$

Here we use the new designations, ω_p (the plasma frequency) and ω_l (the longitudal-mode one), are clear from the formula.

Next, because of each oscillator is contribute to the total polarization, in the linear case (when its internal interactions are neglected), we have

$$P = \sum_j P_j \tag{6}$$

and therefore

$$\varepsilon(\omega) = \sum_j \varepsilon_j(\omega) = 1 + \sum_j \frac{\omega_{pj}^2}{\omega_{0j}^2 - \omega^2} = \prod_j \frac{\omega_{lj}^2 - \omega^2}{\omega_{0j}^2 - \omega^2}. \tag{7}$$

At solid, we should also take into account the crystal lattice potential [3]:

$$\varepsilon(\omega) = \varepsilon_\infty + \sum_j \frac{\omega_{pj}^2}{\omega_{0j}^2 - \omega^2} = \varepsilon_\infty \prod_j \frac{\omega_{lj}^2 - \omega^2}{\omega_{0j}^2 - \omega^2} \tag{8}$$

with ε_∞ is the medium dielectric constant at the very high frequencies.

So, at the second order of smallness (see (1)), the dispersion law for the light in composite is

$$k(\omega) = \frac{\omega}{c} n(\omega) = \frac{\omega}{c}\sqrt{\varepsilon(\omega)\mu} = \frac{\omega}{c}\sqrt{\varepsilon_\infty \prod_j \frac{\omega_{lj}^2 - \omega^2}{\omega_{0j}^2 - \omega^2}\mu}. \tag{9}$$

The unknown parameters of the model, $\{\omega_{lj}, \omega_{0j}\}$, can be obtained experimentally, for example, by the secondary emission spectrum. This way, at the normal incidence, the sample's reflection [4] is

$$R(\omega) = \left|\frac{n(\omega)-1}{n(\omega)+1}\right|^2 = \left|\frac{k(\omega)-\omega/c}{k(\omega)+\omega/c}\right|^2. \tag{10}$$

The analyses of (10) gives the reflection peaks ($R = 1$) at $\omega = \omega_{lj}$ and $\omega = \omega_{0j}$. So, the model parameters, $\{\omega_{lj}, \omega_{0j}\}$, are the phenomenological ones, and these are the edges of the reflection bands in the spectra.

Note, that at the luminescence spectrum, these frequencies are the luminescence peaks because of the zero group speed of light:

$$\mathrm{v}(\omega) = \frac{d\omega}{dk} = \left[\frac{dk(\omega)}{d\omega}\right]^{-1}. \tag{11}$$

Herewith, the effective mass of the light quanta is

$$m(\omega) = \left[\frac{\partial^2 W}{\partial p^2}\right]^{-1} = \hbar\left[\frac{\partial^2 \omega}{\partial k^2}\right]^{-1} = \hbar\left[\mathrm{v}(\omega)\frac{d\mathrm{v}(\omega)}{d\omega}\right]^{-1} \tag{12}$$

and is non-zero to be prospective for the polariton Bose-condensation [5].

IOP Conf. Series: Materials Science and Engineering **859** (2020) 012003 doi:10.1088/1757-899X/859/1/012003

3. Simulation

We simulate the 1D aluminum oxide photonic crystal film synthesized by the aluminum anodizing during the acid etching [6-12]. The film is about 20 μ*m* and made of regular alternated Al_2O_3 layers. All the odd layers are 193 *nm* long and 40% porous, all the even ones are 193 *nm* long and 60% porous. The simulation results are presented at the fig. 1.

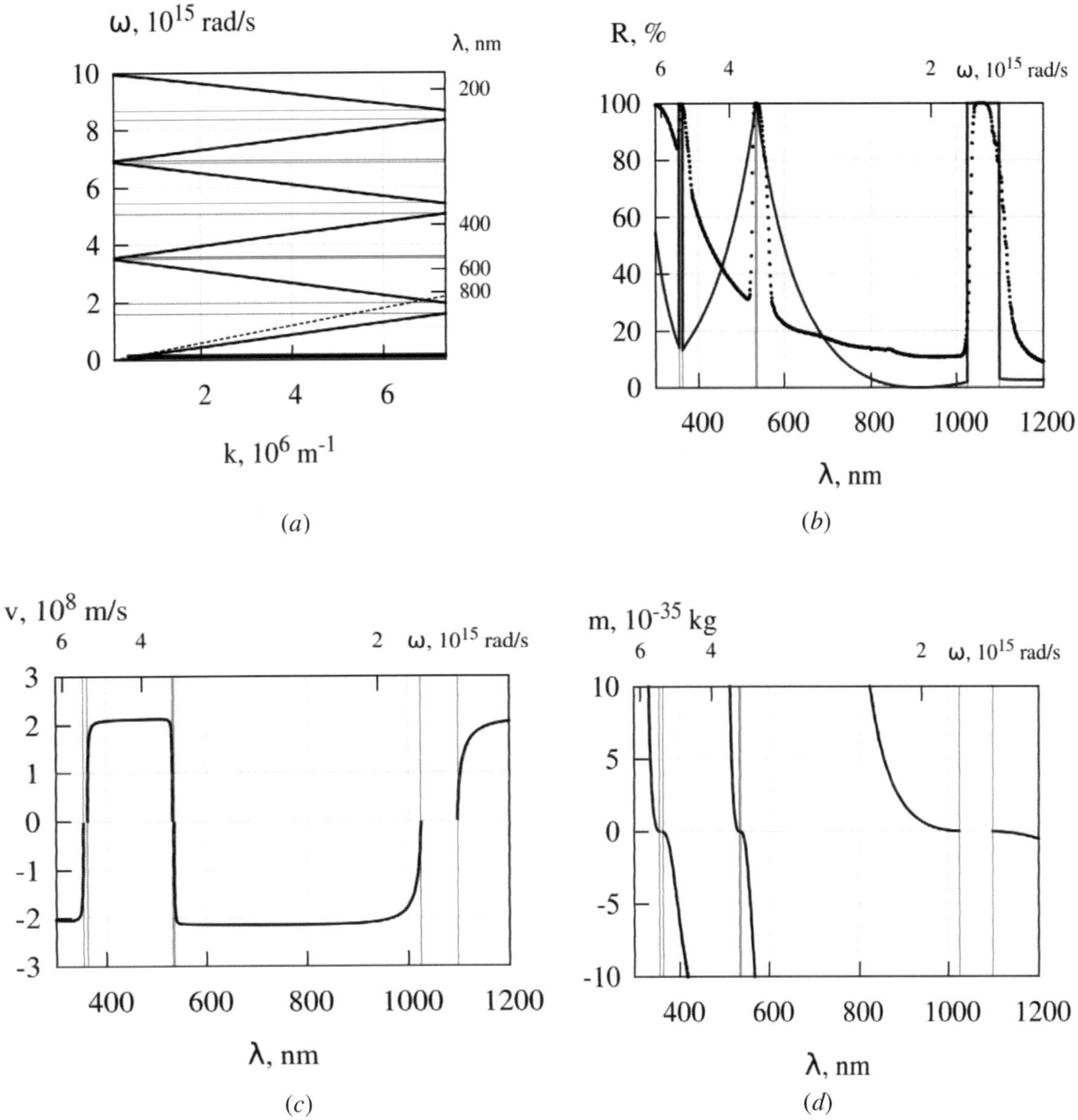

Figure 1. The simulated characteristics of the aluminum oxide photonic crystal: (*a*) the light dispersion law, (*b*) the crystal reflectance (dots are the spectral data), (*c*) the group speed of light in the crystal, and (*d*) the effective mass of the speed quanta. At fig. 1*a*, dotted line marks the dispersion law for the vacuum $\omega = c \times k$. Gray lines mark the bandgaps.

4. Discussion

The simulation (see fig. 1) shows, that at the bandgaps edge, there is the room-temperature Bose-Einstein condensation of polaritons in the photonic crystal. In fact, at the bandgap edges, there is the giant density of states (see fig. 1*c*)

$$f = \frac{dp}{dW} = \frac{dk}{d\omega} = \frac{1}{\mathrm{v}} = (\mathrm{v} \to 0) = \infty. \tag{13}$$

At the same time, the critical temperature of condensation is

$$T_c = \frac{2\pi\hbar^2}{mk_B}\left(\frac{n}{\zeta(3/2)}\right)^{2/3}, \tag{14}$$

where m is the polaritons mass, n is its concentration and constants: reduced Planck constant $\hbar$ = 1.055×10^{-34} *J×s,* Boltzmann constant $k = 1.38\times10^{-23}$ *J/K* and Riemann zeta $\zeta(2/3) = 2.612$, and for $n \sim 10^{25}\, m^{-3}$ (Loschmidt constant) and $m \sim 10^{-35}\, kg$ (see fig. 1*d*), is about $T_C \sim 10^7\, K$. So, at room temperatures, at the bandgap edges, there always is the Bose-Einstein condensation of polaritons.

Let's focus at the effect. Due to the energy flux density

$$S = \mathrm{wv} \tag{15}$$

conservation (w is the energy density, and v is its velocity), in the material medium, S multiplies by

$$\frac{c}{\mathrm{v}} = c\left[\frac{d\omega}{dk}\right]^{-1} = c\frac{dk(\omega)}{d\omega} \tag{16}$$

times (compared to the void). So the high-dispersive media acts act the electromagnetic field amplifiers. This way, the probability of the two-photons process increase resonantly to observe the birth and the decay of the paraphoton, "dark matter" axion-like particle [13].

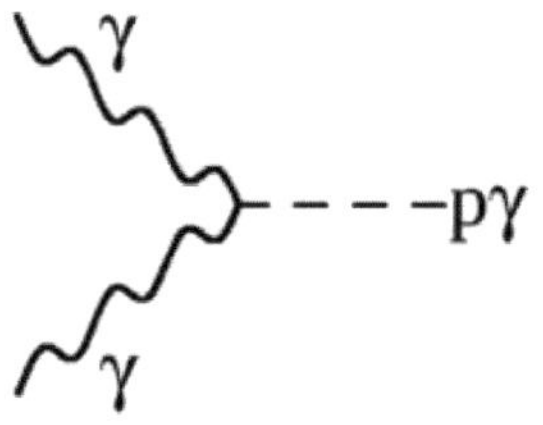

Figure 2. The two photons – to – a paraphoton process.

In the quantum field theory [14] this describes by the equation

$$\nabla^2\phi - \frac{1}{c}\frac{\partial^2\phi}{\partial t^2} = -gE^2, \tag{17}$$

where ϕ is the scalar paraphotonic field, g is the gauge constant.

The solution for (17) is [15]

$$\phi(\vec{r},t) = \frac{1}{4\pi}\int \frac{gE^2(\vec{r}\prime,t)}{|\vec{r}-\vec{r}\prime|}d^3\vec{r}\prime = \frac{gl}{2k}[E_0 e^{i(kx-\omega t+\pi/4)}]^2. \tag{18}$$

Here, $E = E_0 \exp[i(kx - \omega t)]$ is the excite elecromagnetic wave, $l >> k^{-1}$ is the size of the conversion region and the geometry is the one dimensional.

So, the probability of the paraphoton processes is proportional to the energy density (E^2) which is proportional to the density of states (f) and, at the bandgaps, tends to the infinity (see (13)). Therefore, the fig. 2 process becomes resonant and paraphotons generate in the crystal.

IOP Conf. Series: Materials Science and Engineering **859** (2020) 012003 doi:10.1088/1757-899X/859/1/012003

5. Conclusion

The abstract medium oscillators model can deal with not only the normal but with the photonic crystal. This way, we get the correct results (see fig. 1) and can easily modify the model to evaluate the thermo-, the magneto-, the acousto-, the quantum optic- and the other effects. The very intriguing of them is the two photon process to perform the solid-state detection of paraphotons at a laboratory.

Acknowledgements
The work is supported by the Russian Fund for Basic Researches (grant no. 18-02-00181).

References

[1] S M Rytov 1956 *Soviet Physics JETP* 2, 466.
[2] A F J Levi 2016 *The Lorentz oscillator model* in *Essential Classical Mechanics for Device Physics* (Essential Classical Mechanics for Device Physics).
[3] Tatsumi Kurosawa 1961 *Journal of the Physical Society of Japan* 16 1298.
[4] M Born M and E Wolf (1999) *Principles of Optics: Electromagnetic Theory of Propagation, Interference and Diffraction of Light, 7th ed.* (Cambridge University Press).
[5] J Klaers, J Schmitt, F Vewinger and M Weitz 2010 *Nature* 468 545.
[6] M Ashurov, V Gorelik, K Napolskii and S Klimonsky 2019 *Photonic Sensors* DOI:10.1007/s13320-019-0569-2
[7] V S Gorelik, Dongxue Bi, Guang Tao Fei, Shao Hui Xu and Xu Dong Gao 2019 *Inorganic Materials* 55 355.
[8] V S Gorelik, M M Yashin, Dungxue Bi and Guang Tao Fei 2018 *Optics and Spectroscopy* 124 167.
[9] V S Gorelik, P P Sverbil, V V Filatov, Dongxue Bi, Guang Tao Fei and Shao Hui Xu 2018 *Photonics and Nanostructures – Fundamentals and Applications*. 32 6.
[10] V S Gorelik, M M Yashin and A I Vodchits 2017 *Physics of Wave Phenomena* 25 175.
[11] V S Gorelik, Yu P Voinov, V V Shchavlev, Dongxue Bi, Guo Liang Shang and Guang Tao Fei 2017 *Journal of Surface Investigation: X-ray, Synchrotron and Neutron Techniques* 11 246.
[12] V S Gorelik, A V Pudovkin and V V Filatov 2016 *Journal of Russian Laser Research* 37 604.
[13] I S Alimkina, V S Gorelik, V V Filatov and S V Pichkurenko 2019 *Journal of physics: conference series*, in press.
[14] K van Bibber, N R Dagdeviren, S E Koonin 1987 *Physical review letters* 59 759.
[15] R Feynman R, R Leighton, M Sands 1963 *The Feynman lectures on physics*. Vol.2. http://www.feynmanlectures.caltech.edu/info/ (date of access: 05.06.2019).

IOP Conf. Series: Materials Science and Engineering **859** (2020) 012004 doi:10.1088/1757-899X/859/1/012004

Multi-electrode Designed Shape for Small Scale Plasma Incinerator

Grich Kongphet[1, a], Tanakorn Wongwuttanasatian[2, b], Amnart Suksri[3, c]

[1]Department of Electrical Engineering, Faculty of Engineering, Khon Kaen University, Khon Kaen, Thailand
[2,3]Center of Alternative Energy Research and Development, Faculty of Engineering,Khon Kaen University, Khon Kaen, Thailand.

[a]kogrich@kkumail.com, [b]tanwon@kku.ac.th, [c] samnar@kku.ac.th

Abstract. Currently electronic waste has increased due to the growth of an industrial economy. Conventional method of waste destruction was to use burner that heats up and finally forming hydrocarbon to an atmosphere. Recently, plasma technology is an alternative method that has an efficiency and safe to use as a mean for waste destruction due to its higher temperature. This research is an improvement design of electrodes tip used for plasma head assembly using small scaled power supply. The designed shape and materials such as graphite, steel and tungsten were simulated and used to evaluate the generated plasma. In addition, electrode designed group were compared for a current density and current distribution from a simulation which will also determine for a plasma length, temperature and durability of material on actual experimental. Electrode shape was designed with sharp cornered that has cross sectional area indicated that it has highest current distribution at the tip of electrode surface. Based on the simulation results, the proposed designed electrode was capable of highest current density when compared with others. The simulated and experimental results have been shown to agreed well for plasma length and temperature.

1. Introduction

Currently, plasma technology is very effective and efficient for waste destruction. Plasma have high energy and high temperature as the fourth state of matter. Electrodes were designed to have highest current density for plasma torch. High current DC power supply and compressed air were used for creating a plasma generator [1]. Electronic waste nowadays has increased considerably from economic growth. Then, the methods of waste management were to use electronic waste as a landfill [2]. However, as for environmental concerned the landfill can be dangerous to the underground water and toxic soil. Other method of waste destruction can be the used of heat from hydrocarbon fuel burners [3].

Plasma technology was used as an alternative method for waste destruction recently. So, this research investigates the shape of electrode designed that have the contribution to the plasma current density compare with other electrode types based on computer simulation. Then the proposed electrode was used to design and implement a plasma tip generator. The comparison of current density and testing plasma length, temperature and durability of material were studied.

Content from this work may be used under the terms of the Creative Commons Attribution 3.0 licence. Any further distribution of this work must maintain attribution to the author(s) and the title of the work, journal citation and DOI.
Published under licence by IOP Publishing Ltd

2. Material and method

2.1. Materials
Plasma generator has used varied materials for burner electrode. The chosen electrode material must have highest conductivity and highest molten point. This research proposed the use of an electrode material such as graphite, steel and tungsten due to its characteristic properties [4-5]. Electrodes materials will be compared in terms of durability as well as plasma length for a plasma torch incinerator. A computer simulation software, COMSOL was used for this research.

2.2. Electrode design
Plasma generator was studied, and electrodes were designed in various types in the past. This research investigated on multiple electrode group that has the designed with shape, arrangement and are numbered as shown in figure 1. Type A electrode design was sharp [6], type B electrode design was curved [7], type C electrodes design were multiple conductors [8] and type D electrode that have been designed with the tapered shape with cylindrical tip. Therefore, the expectation of cross-section area of the electrode will directly affect the distribution of current density. Arrangement of electrode was to develop from finding current distribution that was suitable for cross-section area at the electrode tip with computer simulation. Using DC supply voltage of 240 V, DC current at 40 A, compressing air at 0.5 bar, electrode gap was 1 mm, steel electrode material was used [9]. Current distribution of each multiple electrode groups was simulated to produce a plasma torch.

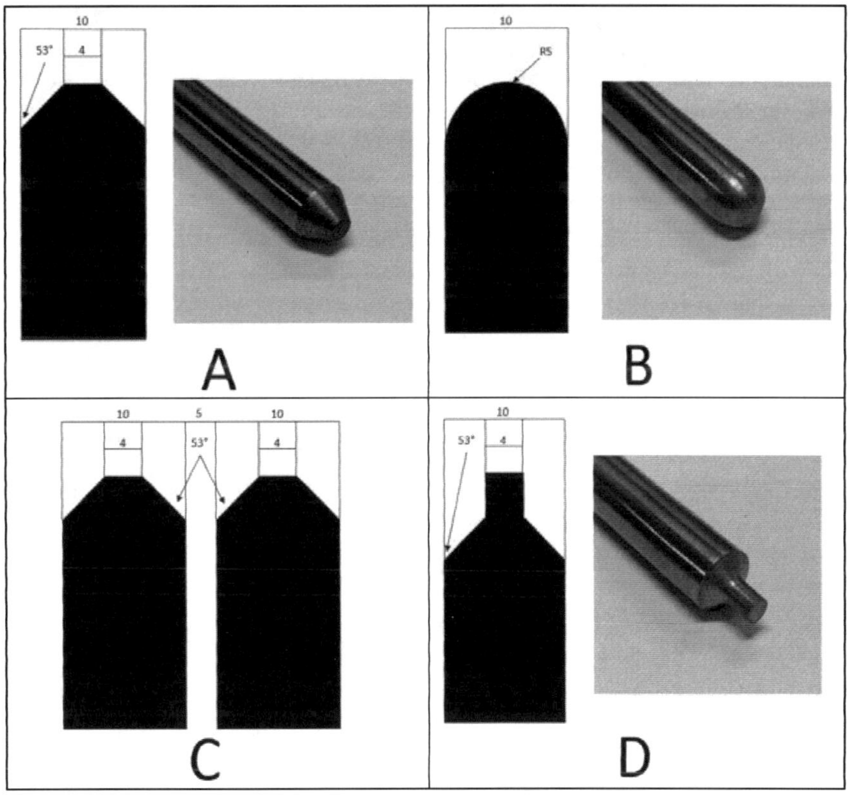

Figure 1. Electrode type A B C and D.

2.3. Implementation of plasma generator
The plasma generator was divided into 3 parts: DC power supply, gas feeder and proposed electrodes as shown in figure 2. The DC power supply voltage of 240 V, compressing was air at 0.5 bar, using air

gap of 1 mm respectively [10]. Materials used are steel, graphite and tungsten were fabricated for a proposed electrode. Furthermore, input current was varied from 15 to 40 A of each test. Then, experimental testing was to compare for a plasma length, temperature and durability of material.

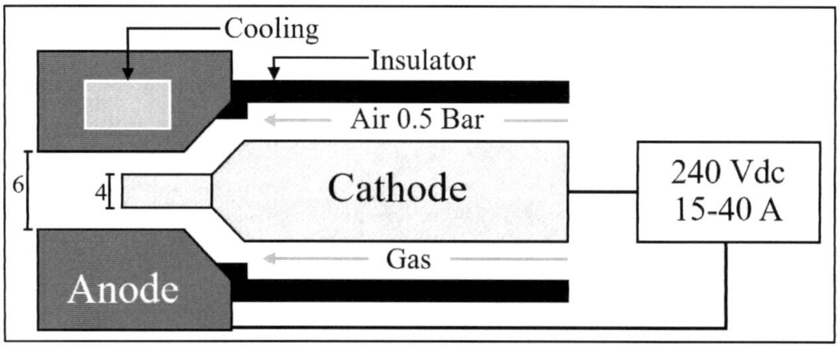

Figure 2. Electrodes, gas feeder and power supply.

3. Results and discussion

3.1. Current density

Table 1 revealed the plasma torch current density at the tip of electrode from a simulation on each multiple electrode group regardless of materials types. The current density distribution indicated into 2 aspects, they are, electrode arrangement and its shape. Electrodes arrangement affected the current density, and electrode with cornered shape that has cross sectional area indicated that it has highest current density distribution at the tip of electrode surface. From experimental results with various electrode type, the current density of type A, B, C and D that were distributed at cross-sectional area of electrode tip were $6.52 \times 10^6, 2.21 \times 10^6, 9.31 \times 10^4$ and $7.39 \times 10^6\ A/m^2$ respectively. Moreover, figure 3 has shown that the shape of the proposed electrode has significantly delivered great magnitude of current distribution. The simulation has also found that electrode design with the cornered shape will deliver highest current density plasma. From figure 3, it was observed that current density of type A and C was distributed at the electrode tip except that type C was multi electrodes. Moreover, current density of type B was distributed at the cross-section area along curve of the electrode tip. The current density of type D was distributed at the cylindrical of the electrode tip and provided the highest value.

Table 1. Current density A, B, C and D.

Types	A-Sharp Electrode(A/m^2)	B-Curved Electrode(A/m^2)	C-Multiple Electrode(A/m^2)	D-Proposed Electrode(A/m^2)
Current density	6.52×10^6	2.21×10^6	9.63×10^4	7.39×10^6

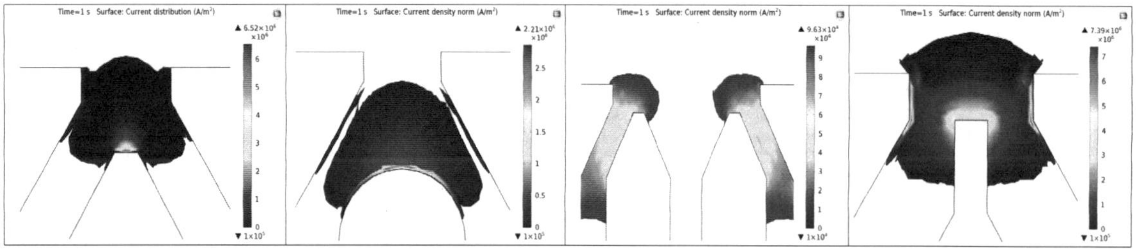

Figure 3. Current distribution of electrode type A, B, C and D.

3.2. Plasma length and temperature

Figure 4 showed the plasma length and temperature of plasma with varied input current from 15 to 40 A. The plasma length and temperature measurement results were obtained from the high-speed video recording. Plasma length was measured by standard ruler that was mounted at the base. For the temperature measurement, the comparison between color temperature shade of plasma torch and international standard shade was used to indicate the plasma temperature. For experiment, the test has varied material including graphite, steel and tungsten for plasma length and temperature investigation. According to the measurement results, there were a clear upward trend in the length and temperature at input current of 40 A that provide the longest length and highest temperature. The results obtained were measured from the length of highest colour temperature. It can be observed that, steel electrode (a) the plasma generated with a length of 37 mm and temperature of 7500 K. As for tungsten electrode (b), the plasma length and temperature were at 27 mm and 7500 K respectively. In case of graphite electrode (c), plasma length obtained was at 21 mm with temperature of 7500 K. It is noticed that the temperature value of each material was relatively close at 7500 K approximately.

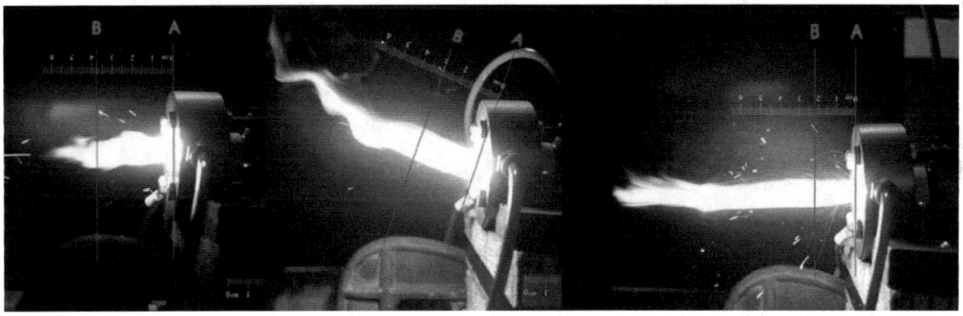

a) steel electrode b) tungsten electrode c) graphite electrode

Figure 4. Plasma torch from various materials.

From figure 4, steel was the material that generated highest plasma temperature due to its lowest melting point. On the other hand, the length of the plasma developed, graphite has been shown to provide longest length of plasma torch among other materials. As for the material durability, the merit order are tungsten, steel and graphite. Further observation was found that electrodes damages were molten as for steel and tungsten case. Graphite electrode, however, the material was brittle and have been found cracked finally. Material selection could be chosen according to the application as well as the compromised between the current density distribution and its durability.

4. Conclusion

In conclusion, the development of the designed electrode for enhancing the plasma torch performance was investigated. From the results, the proposed electrode designed delivered the highest current distribution when compared with other type. It was also found that the current density distribution at the tip of electrode surface were of highest value. Thus, the proposed electrode was chosen and fabricated for plasma generator. Results obtained from actual test and computer simulation have agreed well with each other. It was indicated that plasma incinerator with long plasma length and high temperature with small scale DC power supply could be utilized. Graphite electrode was chosen for an upward trend for the highest temperature plasma torch as well as better current density performance among other materials.

Acknowledgments

This research study is financially supported by the Center for Alternative Energy Research and Development, Khon Kaen University, Thailand.

IOP Conf. Series: Materials Science and Engineering **859** (2020) 012004 doi:10.1088/1757-899X/859/1/012004

References

[1] P. Khongkrapan and N. Tippayawong, "Development of a Thermal Plasma Torch," no. February 2015.

[2] A. Mitrasinovic, L. Pershin, J. Z. Wen, and J. Mostaghimi, "Recovery of Cu and valuable metals from E-waste using thermal plasma treatment," Jom, vol. 63, no. 8, pp. 24–28, 2011.

[3] J. Szalatkiewicz, "Metals recovery from artificial ore in case of printed circuit boards, using plasmatron plasma reactor," Materials (Basel)., vol. 9, no. 8, 2016.

[4] H. Nishikawa, M. Ibe, M. Tanaka, T. Takemoto, and M. Ushio, "Effect of DC steam plasma on gasifying carbonized waste," Vacuum, vol. 80, no. 11–12, pp. 1311–1315, 2006.

[5] D. Shin, B. Swain, C. Han, Y. Kim, C. G. Lee, and K. S. Park, "Comparison of different tungsten precursors for preparation of tungsten nanopowder by RF induction thermal plasma," Int. J. Refract. Met. Hard Mater., vol. 86, no. June 2019, p. 104995, 2020.

[6] S. GHORUI and A. K. DAS, "Arc plasma devices: Evolving mechanical design from numerical simulation," Pramana, vol. 80, no. 4, pp. 685–699, 2013.

[7] D. Zhang, L. Zheng, X. Hu, and H. Zhang, "Numerical studies of arc plasma generation in single cathode and three-cathode plasma torch and its impact on plasma spraying," Int. J. Heat Mass Transf., vol. 98, pp. 508–522, 2016.

[8] S. Mohsenian, M. S. Esmaili, B. Shokri, and M. Ghorbanalilu, "Physical characteristics of twin DC thermal plasma torch applied to polymer waste treatment," J. Electrostat., vol. 76, pp. 231–237, 2015.

[9] A. Leal Vieira Cubas et al., "Inertization of heavy metals present in galvanic sludge by DC thermal plasma," Environ. Sci. Technol., vol. 48, no. 5, pp. 2853–2861, 2014.

[10] H. Nishikawa, M. Ibe, M. Tanaka, T. Takemoto, and M. Ushio, "Effect of DC steam plasma on gasifying carbonized waste," Vacuum, vol. 80, no. 11–12, pp. 1311–1315, 2006.

IOP Conf. Series: Materials Science and Engineering **859** (2020) 012005 doi:10.1088/1757-899X/859/1/012005

Relations between microstructure and hardness of plain carbon steels using eddy current technique

Isadora M. O. A. COSTA[1,2], **Christophe MESPLONT**[2], **Jérémie BOUQUEREL**[1], **Jean-Bernard VOGT**[1*]

[1] Univ. Lille, CNRS, INRAE, Centrale Lille, UMR 8207 - UMET - Unité Matériaux et Transformations, F-59000 Lille, France
[2] Bekaert NV, Bekaertstraat 2, 8550 Zwevegem, Belgium

* jean-bernard.vogt@univ-lille.fr

Abstract. Heat-treatments were carried out on carbon steels ranging from 0.03 to 0.78 wt%C, in order to produce various ferrite-pearlite microstructures. The specimens were characterized by metallographic examinations and microhardness measurements. A clear dependence was found between microstructure characteristics and eddy current outputs measured by means of an electromagnetic sensor: resistance was observed to increase, while inductive reactance decreased in the order of pearlite and ferrite microstructures, and with decreasing interlamellar spacing of pearlite. These components are related to the electrical resistivity and magnetic permeability of the steels. The potentiality of this technique was highlighted for monitoring phase proportions, quantitatively assessing pearlite interlamellar spacing, giving also information about mechanical properties, such as hardness. It reveals the great potential of eddy current testing as a reliable non-destructive tool for metallurgical and mechanical characterization of carbon steels.

1. Introduction

The knowledge and adequate control of the microstructural and mechanical properties of steels are a part of a general approach for improving performance of manufacturing process and enhancing final product quality. In this context, non-destructive techniques have attracted considerable interest for the evaluation of material characteristics and monitoring in-service degradation of components. Among the various non-destructive techniques, eddy current testing is a widely used electromagnetic method for discontinuities and defects detection, corrosion damage inspection and coating thickness measurement in conductive materials [1,2]. More recently, eddy current method has been developed for material characterization, due to its sensitivity to material's permeability and resistivity variations [3-5]. Indeed, magnetic and electrical properties of ferromagnetic materials are known to be altered by their chemical composition [6,7], microstructure [8-11], and to some extent, by their residual stresses [12,13].

In the present work, eddy current method was used to investigate the influence of metallurgical characteristics on the electrical resistivity and magnetic permeability of carbon steels. Heat treatments were performed to produce various ferrite-pearlite microstructures. Microstructural observation and microhardness tests were conducted on the thermally treated carbon steels. Eddy current testing was

Content from this work may be used under the terms of the Creative Commons Attribution 3.0 licence. Any further distribution of this work must maintain attribution to the author(s) and the title of the work, journal citation and DOI.
Published under licence by IOP Publishing Ltd

IOP Conf. Series: Materials Science and Engineering **859** (2020) 012005 doi:10.1088/1757-899X/859/1/012005

carried out using an electromagnetic sensor. The objective of this study is to evaluate the feasibility of eddy current testing for measuring and monitoring carbon steel characteristics.

2. Experimental procedure

2.1. Materials and heat treatments

The samples examined in this study were obtained from carbon steels with carbon content varying from 0.03 to 0.78 wt%C, the compositions of which are given in Table 1. In order to produce specimens with different types of grains, ferrite and pearlite, the steels with various amounts of carbon were austenitized at 950°C and cooled at 580°C. Group A consists of different batches of steels with different ferrite/pearlite fractions. It is worth to mention that heat treatments were carried out at fast cooling rates, which configures a non-equilibrium transformation. As a result, the pearlite volume fraction obtained for each steel composition was higher than it would be obtained in an equilibrium condition. Group B includes entirely pearlitic samples with various lamellar spacings obtained from a 0.78 wt%C steel. Aiming at obtaining different interlamellar spacings, the samples were austenitized at 950°C and cooled at six different temperatures ranging from 540 to 640°C.

2.2. Microstructural and mechanical characterization

Metallographic preparation of the heat-treated steel samples involved abrasive cutting, mounting, followed by grinding and polishing using successively finer water-based diamond suspensions. The microstructural observations were performed by light optical microscopy (LOM) and scanning electron microscopy (SEM). Prior-austenite grain sizes and volume fractions of phases were measured on randomly selected fields in the LOM, using the image analysis software ImageJ/Fiji. Grain sizes were measured by the intercept method after thermal etching with alkaline sodium picrate solution to reveal austenite grain boundaries [14,15], while the quantification of phase fraction was done by phase-contrast image analysis after color etching with 3% Nital and 10% potassium metabisulfite solution [14]. The interlamellar spacing was measured on SEM images using the Underwood method, by which the mean true interlamellar spacing could be randomly estimated according to its intersection procedure [16]. Prior-austenite grain sizes were considered approximately equal for all samples, varying from 20 to 40 μm in diameter. One-way analysis of variance (ANOVA) was carried out to confirm that the differences in the estimated phase fraction and interlamellar spacing values between the samples were statistically significant. Mechanical characteristics were determined via Vickers microhardness measurements.

Table 1. Summary of the chemical composition, various heat treatments, pearlite volume fraction (P) and average interlamellar spacing ($\bar{S}$), and microhardness values of the investigated steels

AISI designation	Chemical composition (wt%)				Heat treatments	P (%)	$\bar{S}$ (nm)	Vickers hardness number
	C	Mn	Si	Fe				
Group A								
1000	0.03	0.15	0.01	Bal.		0	–	112
1020	0.20	0.73	0.26	Bal.		28.7	243 ± 68	170
1035	0.35	0.61	0.21	Bal.		78.3	196 ± 21	209
1050	0.50	0.59	0.21	Bal.	Austenitizing at 950°C	94.7	157 ± 16	266
1055	0.52	0.61	0.21	Bal.	Cooling at 580°C	95.8	156 ± 13	272
1060	0.58	0.61	0.19	Bal.		98.1	153 ± 16	295
1065	0.65	0.61	0.18	Bal.		99.1	155 ± 30	317
1080	0.78	0.59	0.24	Bal.		100	126 ± 20	354
Group B								
1080	0.78	0.59	0.24	Bal.	Cooling at 540°C	100	95 ± 9	374
1080	0.78	0.59	0.24	Bal.	Cooling at 560°C	100	103 ± 12	370

1080	0.78	0.59	0.24	Bal.	Cooling at 580°C	100	126 ± 20	354
1080	0.78	0.59	0.24	Bal.	Cooling at 600°C	100	129 ± 22	350
1080	0.78	0.59	0.24	Bal.	Cooling at 620°C	100	153 ± 27	328
1080	0.78	0.59	0.24	Bal.	Cooling at 640°C	100	185 ± 37	322

2.3. Eddy current testing

Eddy current testing is based on the principles of the electromagnetic induction. It consists of an alternating current generator that applies an alternating voltage to a coil circuit. A current flows through it, which causes an alternating magnetic field to develop around the coil, in phase with the voltage and the primary current. If a conductive material is brought into the proximity to this changing magnetic field, eddy currents are induced into it. Eddy currents cause a secondary flux to develop into the material in opposite polarity to the primary flux. A change in the overall net magnetic flux results in a changing of coil impedance, composed by a resistive component and an inductive term, as indicated:

$$Z = \sqrt{R^2 + X_L^2} \tag{1}$$

where the resistance R and the inductive reactance X_L represent, respectively, the real and imaginary parts of the coil's impedance Z in ohms. Impedance changes depend on the magnitude of eddy currents, which are a function of electrical conductivity and magnetic permeability of the material, beyond parameters such as test frequency and fill-factor. Thus, eddy current testing can be assumed as an indirectly response of microstructure characteristics of conductive materials. In this study, conventional coil arrangement connected through an oscillator circuit and an amplifier to a data acquisition computer was employed. Tests were performed at room temperature and at a fixed test frequency, in the way that electrical resistivity and magnetic permeability were the only effective parameters affecting the material response to the induced current. The impedance plane is the convention for representing eddy current signals, and voltage signal variations were used to display impedance variations of the coil.

3. Results and discussions

3.1. Ferrite-pearlite structures: effect of carbon content

Typical microstructures of the heat-treated steel samples are shown in optical micrographs of Figure 2. The ultra-low carbon 1000 steel was used to produce a fully ferritic microstructure of α-phase, which is a solid solution of carbon in iron. Hypoeutectoid steels were used for producing ferrite-pearlite structures with pearlite fractions varying from 28.7 to 99.1%. In these structures, a network of proeutectoid ferrite (bright regions) is observed along the grain boundaries surrounding areas of pearlite (dark regions), which is a constituent consisted of alternate lamellae of eutectoid α ferrite and cementite (Fe_3C) phases. The eutectoid 1080 steel exhibited a fully pearlitic microstructure.

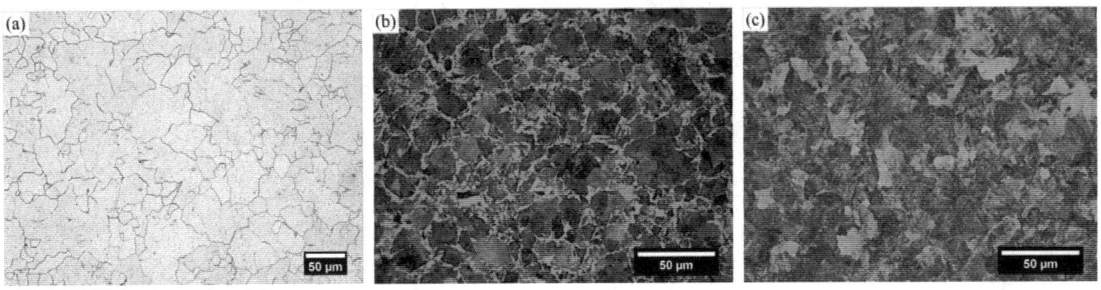

Figure 1. Optical micrographs of (a) ferrite, (b) ferrite-pearlite and (c) pearlite structures

Hardness was observed to increase linearly with the increase of carbon content. As the pearlite fraction increases, the proportion of cementite, which is a relatively hard phase, also increases and the hardness of the steel rises accordingly. Figure 2-a shows the impedance changes with respect to the Vickers hardness values of the ferrite-pearlite microstructures. Impedance values decreased with the increase of carbon content, and so with the increase of the Vickers hardness. The impedance plane that includes the resistance and the inductive reactance components is shown in Figure 2-b. Inductive reactance values were observed to decrease, while resistance increased with increasing carbon content.

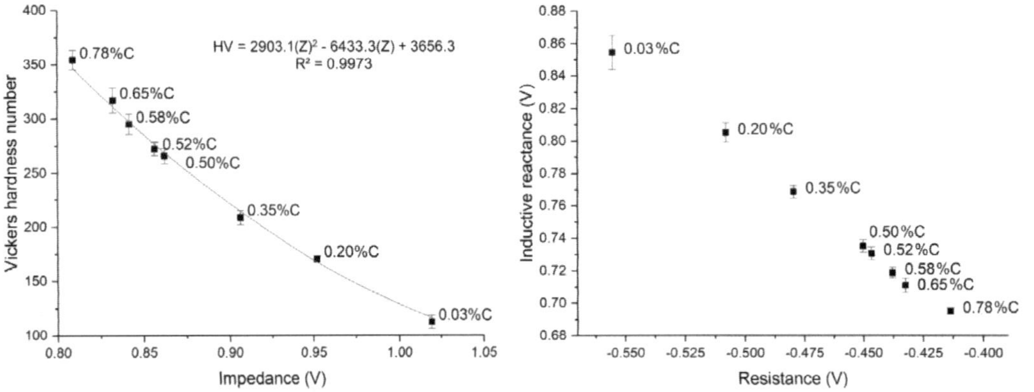

Figure 2. Relations between hardness and eddy current outputs (left); (b) impedance plane of plain carbon steels (right)

3.2. *Pearlite structures: influence of pearlite interlamellar spacing*

Aiming at evaluating more clearly the role of the spacing of pearlite lamellae, fully pearlitic samples were produced by varying the temperature at which austenite is transformed. A decrease in the transformation temperature was observed to refine the interlamellar spacing between ferrite and cementite lamellae, as observed in Figure 3. In the SEM images, the dark regions are the α-phase that occurs both in the pearlite and in the simple ferrite grains, while the Fe_3C cementite is white and sticks out of the ferrite matrix. Narrower $\bar{S}$ means more interfaces between ferrite matrix and cementite lamellae, leading to an increase in hardness that correlated well with the decrease in impedance values observed in Figure 4-a. In Figure 4-b, the complex impedance plane shows that, inductive reactance values decreased with decreasing interlamellar spacing of pearlite, while resistance values increased.

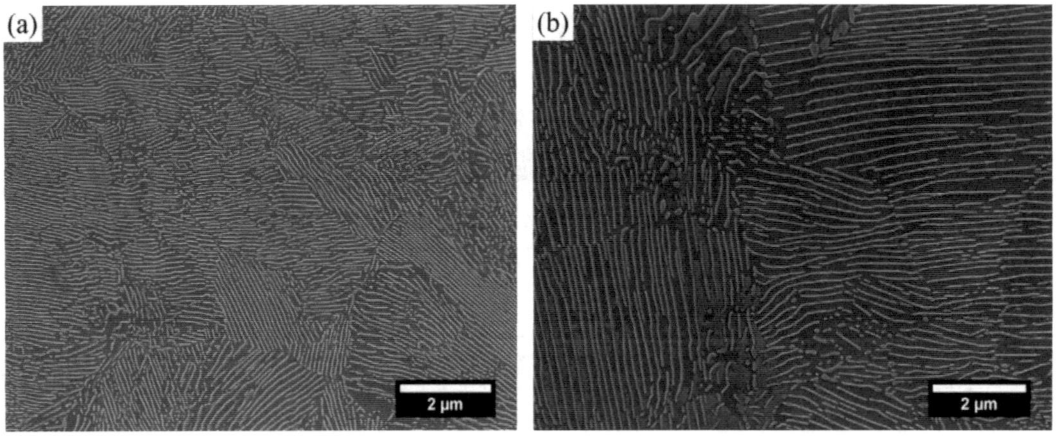

Figure 3. SEM images of 1080 steels cooled at (a) 540°C ($\bar{S}$=95 nm), and (b) 640°C ($\bar{S}$=185 nm)

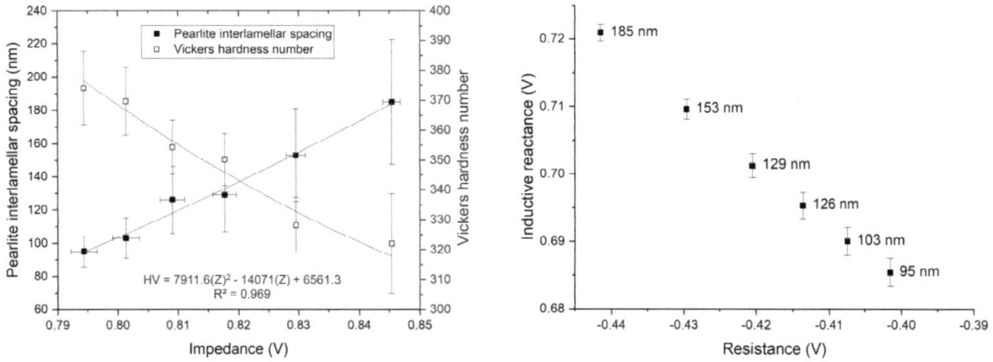

Figure 4. Relations between interlamellar spacing, hardness and eddy current outputs (left); impedance plane of eutectoid steels (right)

3.3. Assessment of the effect of microstructure on physical properties of carbon steels

The potentiality of eddy current testing was highlighted to distinguish the different arrangements of ferrite-pearlite microstructures and estimate pearlite interlamellar spacing, giving also information about hardness of thermally treated carbon steels. Carbon steels could be characterized by measuring changes in inductive reactance and effective resistance of an electromagnetic sensor operating at low frequencies. These components are related, respectively, to the magnetic permeability and electrical resistivity of the ferromagnetic steels [1,2]. The overall electromagnetic behavior in a multiphase structure was observed to be a function of their different microstructural features. The resistance increased while inductive reactance decreased with carbon content due to the different amounts of pearlite in the steels. It was also observed an increase of resistance and a decrease of inductive reactance with the decrease of pearlite interlamellar spacing. These effects were attributed to the propagation and nucleation processes of magnetic domain walls, and to the electrons mobility into the crystalline microstructure. The presence of a lamellar carbide network influenced the magnetic response between the various microstructures, as it represents effective pinning sites that hinder domain wall motion [4,17-19]. Moreover, cementite layers can act as scattering centers for electrons, which would lead to the increase of steel resistivity [11]. For each material, the mechanically harder specimens exhibited higher resistance and lower inductive reactance, indicating that hardness can be readily measurable by eddy current testing.

4. Conclusions

From the study to evaluate the influence of microstructure and hardness on the electrical resistivity and magnetic permeability of carbon steels by means of eddy current measurements, it can be concluded:

1. Resistance (electrical resistivity) was observed to increase while inductive reactance (magnetic permeability) decreased in the order of pearlite, ferrite-pearlite and ferrite microstructures, and with a narrower interlamellar spacing of fully pearlitic microstructures.
2. Quantitative linear correlations were obtained between microstructure features, hardness and eddy current outputs. Hence, the results of this research reveal the great potential of eddy current technique to be used as a reliable non-destructive tool for characterization of carbon steels.

References

[1] C.J. Hellier, Handbook of nondestructive evaluation, second ed., McGraw-Hill, New York, 2003
[2] J. García-Martín, J. Gómez-Gil, E. Vázquez-Sánchez, Sensors. 11 (2011) 2525–2565.
[3] M.K. Devine, JOM. 44 (1992) 24–30.
[4] O. Saquet, J. Chicois, A. Vincent, Mater. Sci. Eng. A. 269 (1999) 73–82.

[5] C. Davis, M. Strangwood, A. Peyton, Ironmaking Steelmaking. 38 (2011) 510–517.
[6] D. Mercier, J. Lesage, X. Decoopman, D. Chicot, NDT&E INT. 39 (2006) 652–660.
[7] M.S. Amiri, M. Kashefi, NDT&E INT. 42 (2009) 618–621.
[8] F. Rumiche, J.E. Indacochea, M.L. Wang, JMEP. 17 (2008) 586–593.
[9] S. Konoplyuk, NDT&E INT. 43 (2010) 360–364.
[10] S. Ghanei, M. Kashefi, M. Mazinani, Mater. Des. 50 (2013) 491–496.
[11] L. Zhou, J. Liu, X.J. Hao, M. Strangwood, A.J. Peyton, C.L. Davis, NDT&E INT. 67 (2014) 31–35.
[12] W. Ricken, H.C. Schoenekess, W.J. Becker, Sens. Actuator A-Phys. 129 (2006) 80–85.
[13] B. Cao, T. Iwamoto, Steel Res. Int. 87 (2017) 1–10.
[14] L.B. Bramfitt, O.A. Benscoter, Metallographer's guide: Practices and procedures for irons and steels, ASM International, 2002.
[15] ASTM E112-13 Standard test methods for determining average grain size, ASTM International, West Conshohocken, PA, 2013.
[16] G.F. Vander Voort, A. Róosz, Metallography. 17 (1984) 1–17.
[17] L. Clapham, C. Jagadish, D.L. Atherton, Acta Metall. Mater. 39 (1991) 1555–1562.
[18] S.M. Thompson, B.K. Tanner, J. Magn. Magn. Mater. 123 (1993) 283–298.
[19] J.W. Byeon, S.I. Kwun, Mater. Trans. 44 (2003) 2184–2190.

IOP Conf. Series: Materials Science and Engineering **859** (2020) 012006 doi:10.1088/1757-899X/859/1/012006

Analytical Description of the Kinetic Processes of Forming the Properties of Composites

Irina Garkina, Alexander Danilov

Penza State University of Architecture and Construction, German Titov str., 28, Penza, 440028, Russia

i.a.naum@mail.ru

Abstract. Effective methods are proposed for approximating the kinetic processes of formation of physical and mechanical characteristics of composite materials for solving problems of both single-criterion and multi-criteria optimization of their structure and properties. The approximation of kinetic processes on a given interval with the definition of partial intervals of the maximum length approximation is considered in detail. An approximation algorithm on cubic splines is given. To determine the transfer functions of composites, as dynamic systems with many inputs and outputs, it is recommended to use Laguerre polynomials. The possibility of a cognitive approach is also not excluded with the direct selection of the form of the approximating function.

1. Introduction

The de development of new composite materials with a regulated structure and properties is directly related to the analytical description of the kinetic processes of the formation of their physico-mechanical characteristics.

Often the process of formation of the operational value of the parameter (property) has an exponential aperiodic character (Fig. 1).

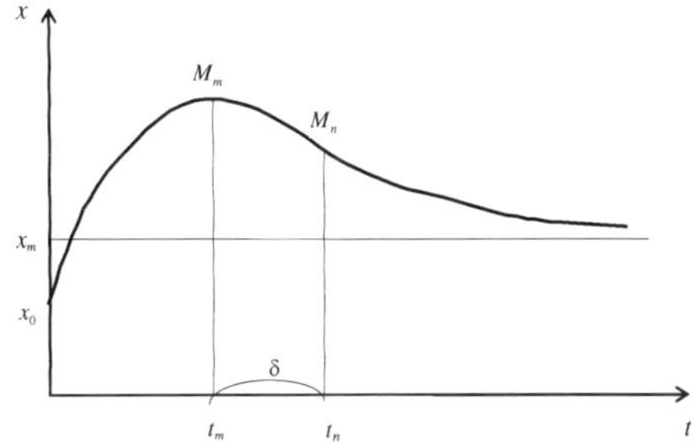

Figure 1. Aperiodic kinetic process.

2. An approximation algorithm with the maximum length of partial

Among a number of methods of approximation of these processes the special place owing to his simplicity is occupied by piecewise and linear approximation, as a rule, on sites of identical length. However, in many cases it is advisable to break down the kinetic processes into separate sections,

Content from this work may be used under the terms of the Creative Commons Attribution 3.0 licence. Any further distribution of this work must maintain attribution to the author(s) and the title of the work, journal citation and DOI.
Published under licence by IOP Publishing Ltd

based on ensuring the maximum length of the approximation sections. As parameters of kinetic processes, various characteristics can be used that integrally determine the processes of formation of properties [1-4].

Data input. Specifies interval $[a,b]$, the relative error of approximation ε_0 in percent. Function $f(x)$ is realized as a program for calculating its values at any point $x \in [a,b]$ or at least at points $x_j, j = \overline{0,N}; x_0 = a, x_N = b$ such that $\left|f_j - f_{j-1}\right| \le \frac{2\varepsilon_0}{1000} f_{cp}$; $f_{cp} = \frac{1}{N}\sum_{j=1}^{N}\left|f_j\right|$, $f_j = f(x_j)$.

The implementation of the algorithm was carried out according to tabular values of $\{x_j, f_j\}_{j=0}^{N}$ for an absolute error of $\varepsilon = \frac{\varepsilon_0}{100}\frac{1}{N}\sum_{j=1}^{N}\left|f_j\right|$ with a tabulation density of $\left|f_j - f_{j-1}\right| \le 0{,}2\varepsilon$.

Tabulation. Here the program for obtaining information is implemented. The following algorithm was used. For $h = 0{,}04(b-a)$, $N = \frac{b-a}{h}$ we compute:

$$x_j = a + jh,;\quad f_j = f(x_j), j = \overline{0,N};\ \varepsilon_h = \max_{1\le j\le N}\left|f_j - f_{j-1}\right|;\ f_{cp} = \frac{1}{N}\sum_{j=1}^{N}\left|f_j\right|, \varepsilon = \frac{\varepsilon_0}{100} f_{cp}.$$

At $\varepsilon_h > 0{,}2\varepsilon$ the value of was decreased by a factor of h.

Piecewise-linear approximation. The approximating function $\tilde{f}(x)$ was represented in the form

$\tilde{f}(x) = A_r + K_r(x - z_r), z_r \le x \le z_{r+1}, r = \overline{0, N_a - 1}$,

at what $z_0 = x_0, z_{N_a} = x_N, \tilde{f}(z_r) = f(z_r)$; $\left|\tilde{f}(x) - f(x)\right| \le \varepsilon$ for $x \in [a,b]$.

Function $\tilde{f}(x)$ is defined by the table $\{z_r, A_r, K_r\}_{r=0}^{N_a}$; z_r - nodes, N_a - number of nodes, $H_r = z_{r+1} - z_r$ - approximation intervals, K_r - angular coefficients, $z_0 = x_0$, $A_0 = f_0$, $A_{r+1} = A_r + K_r H_r$ (continuity $\tilde{f}(x)$).

Thus, the piecewise linear approximation is determined by the parameters K_r, H_r, $r \ge 0$. The maximum of H_r is determined by the choice of point $z_{r+1}, r \ge 0$, as far removed from z_r; $\left|\tilde{f}(x) - f(x)\right| \le \varepsilon, x \in [z_r, z_{r+1}]$; $\tilde{f}(x) = f(x), \tilde{f}(z_{r+1}) = f(z_{r+1})$.

Calculation z_r, A_r, K_r. Accepted $z_0 = x_0, A_0 = f_0$. The calculated values of z_r, A_r are z_{r+1}, $H_r = z_{r+1} - z_r$, $A_{r+1} = f(z_{r+1})$, $K_r = \frac{A_{r+1} - A_r}{H_r}$ (point z_{r+1} is one of the tabulation points x_j).

The algorithm for piecewise linear approximation of the kinetic process $f(x)$ with intervals of maximum length on segment $[a,b]$ is proposed below.

The calculation algorithm z_{r+1} is reduced to the choice of the tabulation point $x_j > z_r$, taking into account condition

$$\max_{j_r < i < j}\left|f_i - A_r - \frac{f_j - A_r}{x_j - z_r}(x_i - z_r)\right| \le \varepsilon; \tag{1}$$

j_r - the number of the tabulation point x_{j_r}, which corresponds to $z_r(z_r = x_{j_r})$.

As soon as for some j the condition (1) is violated, we store x_j as x_j (the number is remembered as x_j) and $K_r^0 = \frac{f_j - A_r}{x_j - z_r}, A_{r+1}^0 = f_j$ is accepted. Then condition (1) for $x_j > z_{r+1}^0$ is checked.

If $\forall x_j \in \left(z_{r+1}^0, x_N\right]$ condition (1) is not satisfied, is accepted $z_{r+1} = z_{r+1}^0, A_{r+1} = A_{r+1}^0, K_r = K_r^0$, $j_{r+1} = j_{r+1}^0$ and proceed to the calculation of z_{r+2}.

If condition (1) is satisfied for some $x_j \in \left(z_{r+1}^0, x_N\right]$, then the transition to the previous stage is performed (r does not increase).

This will define all the triads $\{z_r, A_r, K_r\}_{r=0}^{N_a-1}$. In the last triad $\{z_{N_a}, A_{N_a}, K_{N_a}\}$, it suffices to compute $z_{N_a} = z_{r+1} = x_N, K_r = K_{N_a-1}$.

Note that if $f(x)$ satisfies condition $|f(x)| \geq \alpha \; \forall x \in [a,b]$, then instead of condition (1) (using the averaged relative error $\bar{\varepsilon}_o = \dfrac{|\tilde{f} - f|}{f_{cp}} 100\%$), one can also use condition

$$\max_{j_r \leq i \leq j} \left| f_i - A_r - \frac{f_j - A_r}{x_j - z_r}(x_i - z_r) \right| \leq \frac{\varepsilon_0}{100} |f_i|.$$

If, when condition (1) $x_j = x_N$ is satisfied, then $z_{N_a} = z_{r+1} = x_N, K_r = K_{N_a-1} = \dfrac{f_N - A_r}{x_N - z_r}$; the calculations are completed.

3. Approximation by cubic splines

We also give another possible version of the approximation of the kinetic process, based on the use of interpolation by polynomials in individual sections, and not on the basis of constructing a global interpolation polynomial over the whole interval. The smooth piecewise polynomial functions (composed of polynomials of the same degree) obtained in this case are splines. We confine ourselves to cubic splines (composed of polynomials of the third degree). We propose an algorithm for constructing cubic splines of class C^2 (twice continuously differentiable functions).

Here it is assumed that the values of the function $f(x)$: $f_i = f(x_i), i = \overline{1,N}$. Are given in grid $[a,b]$ at grid nodes $\omega\{a = x_1 < x_2 <, \mathrm{K}, x_N = b\}$. According to the preceding, the interpolating cubic spline $S(x)$ must satisfy the following conditions:

$$S(x_i) = f_i,\; S'(x_i) = f(x_i), S''(x_i) = f'(x_i). \tag{2}$$

The cubic spline $S(x)$ on each of the segments $[x_i, x_{i+1}]$ is determined by four coefficients; for its construction on the whole interval $[a,b]$ it is required to determine $4N$ coefficients. From the condition that the spline belongs to the class C^2, it is assumed that all internal interpolation points $x_i, i = \overline{1, N-1}$, not only the spline $S(x)$, but its derivatives $S'(x)$ and $S''(x)$ are continuous. From these conditions we obtain $3(N-1)$ equations for determining the unknown spline coefficients. Adding $N+1$ equation from (2), we have $4N-2$ equations; we obtain two missing equations from the boundary conditions (restrictions on the values of the spline and its derivatives at the ends of the interval). The following boundary conditions are most commonly used:

I. $S'(a) = f(a), S'(b) = f(b)$.

II. $S''(a) = f'(a), S''(b) = f'(b)$

III. $S^k(a) = S^k(b), k = 1,2$

IV. $S'''(x_p+0)=S'''(x_p-0),\ p=\overline{1,N-1}$

(the so-called "natural" conditions have the form: $S'(a)=0,\ S''(b)=0$).

At each of the intervals $[x_i,x_{i+1}]$ the spline is not represented in general form

$$S(x)=a_i+b_ix+c_ix^2+d_ix^3,$$

but in some special, allowing to reduce the number of unknown spline coefficients. For this we introduce the notation: $S'(x_i)=m_i, i=\overline{0,N}$; $h_i=x_{i+1}-x_i$, $t=\dfrac{x-x_i}{h_i}$.

On segment $[x_i,x_{i+1}]$ the cubic spline can be written in the form

$$S(x)=f_i(1-t)^2(1+2t)+f_{i+1}t^2(3-2t)+m_ih_it(1-t)^2-m_{i+1}t^2(1+t)h_i.$$

On each of the intervals $[x_i,x_{i+1}]$ it is continuous along with its first derivative everywhere on $[a,b]$. We choose parameters in such a way that the second derivative is also continuous at all internal nodes. We obtain the following system of equations:

$$\lambda_i m_{i-1}+2m_i+\mu_i m_{i+1}=3\left(\mu_i\frac{f_{i+1}-f_i}{h_i}+\lambda_i\frac{f_i-f_{i-1}}{h_{i-1}}\right), \quad (3)$$

Where $\mu_i=\dfrac{h_{i-1}}{h_{i-1}+h_i}$, $\lambda_i=1-\mu_i=\dfrac{h_i}{h_{i-1}+h_i}$, $i=\overline{0,N}$.

Equations (3) should be supplemented by equations

$$2m_0+m_1=3\frac{f_1-f_0}{h_0},\quad m_{N-1}+2m_N=3\frac{f_N-f_{N-1}}{h_{N-1}},$$

obtained from the boundary conditions.

The problem of constructing a cubic spline has been reduced to solving a linear system for the unknown coefficients m_i:

$$2m_0+m_1=3\frac{f_1-f_0}{h_0},\quad \lambda_i m_{i-1}+2m_i+\mu_i m_{i+1}=3\left(\mu_i\frac{f_{i+1}-f_i}{h_i}+\lambda_i\frac{f_i-f_{i-1}}{h_{i-1}}\right),$$

$$m_{N-1}+2m_N=3\frac{f_N-f_{N-1}}{h_{N-1}}. \quad (4)$$

In solving practical problems, system (4) was solved taking into account tridiagonality of the matrix of the system.

4. Approximation by special functions

When solving a whole series of problems in the study of composite materials as nonstationary dynamical systems, it becomes necessary to approximate not only the input and output signals (the "black box"), but also the determination of particular transfer functions (the parameters of coupling the output signals to the input signals). We used for this the decomposition of the kinetic process over various orthonormal systems of special functions, the preference was given to Laguerre polynomials [5]: $L_n(t)=\sum_{\nu=0}^{n}\dfrac{n!(-t)^\nu}{(\nu!)^2(n-\nu)!}$; $\int_0^\infty L_n(t)L_m(t)e^{-t}dt=\begin{cases}0 & at\quad m\neq n\\ 1 & at\quad m=n\end{cases}$,

since in this case the transfer function $W(p)$ is easily defined as a rational function $Y(p)/X(p)$ with simple rules for calculating the coefficients.

In solving practical problems, the human factor also plays an important role: the choice of the form of the approximating function is often determined by the intuition of the experimenter. Thus, according to Table 1, we determined the analytical dependence of the viscosity B of the epoxy composite on the temperature $t, {}^{o}C$, and the percentage content $x,\%$ of a special additive.

Table 1. Experimental data

x \ t	50	55	60	65	70	75	80	85	90	95	100
0	172	119	84	65	50	40	30	25	22	20	19
1	210	145	105	80	65	52	43	35	30	27	25
5	250	160	115	90	75	60	48	40	33	30	28
10	97	65	45	33	22	17	13	10	11	9	8
15	132	91	62	47	36	30	21	18	17	15	12

Proceeding from the cognitive analysis [6,7], the approximating function was given in the form $B = B(x,t)$. Function $B = B(x, t = const)$ was represented as a polynomial of the third degree, and $B = B(x = const, t)$ - functions of the form $y = ae^{-kt}$, $k = k(x)$ for different degrees of the polynomial. Based on these results, an approximation was obtained

$$B = \left(c_0 + c_1 x + c_2 x^2 + c_3 x^3\right)e^{r(x)(t-50)},\ \ 50 \le t \le 100.$$

5. Conclusion

The methods of approximation of the kinetic processes of the formation of physicomechanical characteristics are proposed for synthesis of composite materials with a regulated structure and properties:

- approximation of kinetic processes with the possibility of selecting the maximum length of partial intervals;
- algorithm of approximation by cubic splines;
- applications of Laguerre polynomials to the definition of interelement bonds in composite materials, as complex dynamic systems.

The proposed methods have been tested in the development of a number of special-purpose composite materials.

References

[1] Garkina I.: Modeling of kinetic processes in composite materials. Contemporary Engineering Sciences. 9(8) (2015) 421-425.

[2] Irina Garkina and Alexander Danilov: Mathematical Methods of System Analysis in Construction Materials.IOP Conf. Series: Materials Science and Engineering. 245 (2017) 062014.

[3] Irina Garkina, Alexander Danilov: Analytical design of building materials. Journal of Basic and Applied Research International. 18(2) (2016) 95-99.

[4] Garkina, A. Danilov, Y. Skachkov: Modeling of Building Materials as Complex Systems, Key Engineering Materials. 730 (2017) 412-417.

[5] G. Beytmen, A. Erdeyi: The highest transcendental functions. M.: Science.Vol. 2 (1973) 297 p.

[6] Garkina I., Danilov A.: Tasks of Building Materials from the Viewpoint of Control Theory, Key Engineering Materials. 737(2017) 578-582

[7] Garkina I.A., Danilov A.M.: Experience of development of composite materials: some aspects of mathematical modeling. News of higher educational institutions. Construction. 8 (656) (2013) 28-33.

IOP Conf. Series: Materials Science and Engineering **859** (2020) 012007 doi:10.1088/1757-899X/859/1/012007

Tuning Polymerization Rate Program Block Copolymer Assemblies in PISA: A Simulation Study

Junfeng Wang

School of Materials Science and Engineering, China University of Petroleum (East China), Qingdao, Shandong 266580, People's Republic of China.

wjfgmj@gmail.com

Abstract. As a powerful tool of controlled preparation of polymeric nanomaterials, polymerization-induced self-assembly (PISA), has been developed and extensively studied in the past decade. In this work, the RAFT polymerization system of P4VP as chain transfer agent and St as monomer in methanol was modelled and their PISA process was simulated by dissipative particle dynamics. Different morphologies of P4VP-b-PS was obtained at tuning polymerization rate. The morphology evolution at high conversation was investigated in detail and it was found that the appearance of worm-like micelles was a key point to determine the self-assemblies was vesicles or tubes. In addition, the PISA behavior at low conversion was also studied. The final morphologies was different with that at 100% conversion, but these results were well agreed with experiments. This simulation studies identified polymerization rate as the key factors on tuning morphology in PISA and broadens its research method.

1. Introduction

Polymeric nano-objects have gained great interest due to their versatile applications in self-healing, nanocatalyst carrier, nanoparticles or drug controlled release, and cancer treatment and diagnosis[1]. In preparation of these nano-objects, polymerization induced self-assembly (PISA) has shown great advantages over conventional self-assembly in the selective solution (SA) by virtue of uniform morphology, one-pot facial synthesis route and high solids, which makes the scale-up production of polymeric nano-materials possible[2].

In the past decade, some strategies have been developed to tune the morphologies in PISA. For example, Wan et al tuned block length and feed ratio to synthesize a series of assemblies of poly(4-vinylpyridine)-b-polystyrene (P4VP-b-PS), such as sphere, rod, worm, vesicle and yolk/shell structure[3]. Armes group has made great contribution to the morphology control in PISA, the spherical micelle, worm-like micelle, vesicle and oligolamellar vesicle of poly(ethyleneglycol)-b-poly(2-hydroxypropyl methacrylate) diblock copolymer were successfully achieved though regulating on degree of polymerization (DP)[4]. In addition, unique jellyfish and framboidal vesicle were yielded by using poly(glycerol monomethacrylate)-b-poly(2-hydroxypropyl methacrylate) and poly(glycerol mono-methacrylate)-b-poly(2-hydroxypropyl methacrylate)-b-poly(benzyl methacrylate)[5]. Zhang et al prepared compartmentalized vesicle by employing the blends of AB/BAB block copolymer[6]. An et al demonstrated that star polymer architecture could promote the morphology transformation in PISA[7]. In short, PISA developed from SA has shown many efficient strategies to control morphology, but the in-depth understanding of underlying growth origin like we do in self-assembly still needs more efforts.

Content from this work may be used under the terms of the Creative Commons Attribution 3.0 licence. Any further distribution of this work must maintain attribution to the author(s) and the title of the work, journal citation and DOI.
Published under licence by IOP Publishing Ltd

Compared with SA, the major difference in PISA is the self-assembly happens along with a simultaneous polymerization. Hence, polymerization rate (Pr) plays an important role in the PISA behavior. One example is that different morphologies, vesicles in PISA and nanotubes in SA were prepared using P4VP-b-PS in methanol[8], which suggests that the polymerization in PISA affects the self-assemblies morphology. So, processing an in-depth discussion on the dynamic self-assembly process and uncover the roles of Pr on the morphology regulation has great promise for the development of PISA, especially to develop a feasible route to tune morphology of self-assemblies.

Here, employing the dissipative particle dynamic (DPD) simulation and reaction model[9], which was first used by Lin et al to study the PISA behavior[10], we designed a series of simulations to investigate the effects of Pr on self-assembly morphology.

In this work, the PISA behavior in RAFT polymerization of styrene (St) in methanol with P4VP as the macromolecular chain transfer agent (macro-CTA) was examined. In all simulations, initial concentration of P4VP and St kept constant (1.5 w/w% for P4VP, 10.5w/w% for St). To probe the Pr effect, the reactive probability of bonding between monomers and active end of P4VP denotes Pr, which was set to 0.00001, 0.000025, 0.0001, 0.0002, 0.0004, respectively. Notably, the morphology evolution at different Pr was exhibited when the conversion was almost 100%. Besides that, PISA processes at lower conversion (~ 64%) were also studied, which is closer to the experimental condition. Our study suggested that Pr was one primary factor dominating the self-assembly behavior, and provided a versatile approach to tune self-assembly morphologies for definite functionalized applications.

2. Results and discussion

Figure 1 (a)-(e) showed self-assemblies obtained from PISA at different P_r. With increase of P_r from 0.00001 to 0.000025, 0.0001, 0.0002, 0.0004, lamellar vesicle, vesicle, peanuts-like vesicle and nanotube were produced, respectively. In Figure 1 (f)-(j), P4VP-*b*-PS with various chain lengths in self-assemblies were shown by different colors. And it was obvious that the distribution of the copolymers with various lengths on the five micelles was uniform. In simulations, all monomer conversations achieved 100% at five P_r, and the total polymerization time was $1.3 \times 10^7\tau$ at $P_r = 0.00001$, $5 \times 10^6\tau$ at $P_r = 0.000025$, $1.3 \times 10^6\tau$ at $P_r = 0.0001$, $6.5 \times 10^5\tau$ at $P_r = 0.0002$, $3 \times 10^5\tau$ at $P_r = 0.0004$. Figure 2 (a) showed $\ln(M_0/M)$ and conversion versus normalized time in PISA at five P_r. These curves have similar variation tendency, and indicate that polymerization behavior at different P_r are identical, except that the polymerization time is expanded at lower P_r. On the other hand, from Figure 2 (b), the length distribution of copolymers was almost identical at five different P_r. So in this case, the versatile morphologies obtained at five different P_r was not due to the polymerization behavior or copolymer composition, it might be ascribed to different growth rate of chain, which was further confirmed as follow.

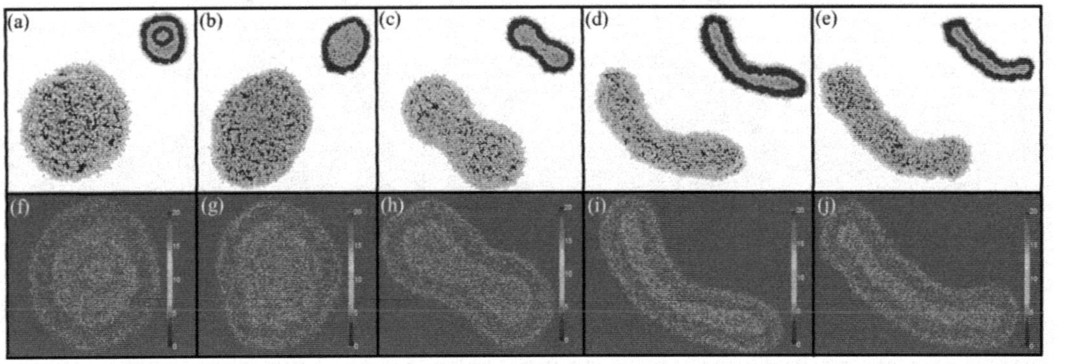

Figure 1. Self-assembly morphologies obtained at varied Pr, (a) Pr = 0.00001 (b) Pr = 0.000025 (c) Pr = 0.0001 (d) Pr = 0.0002 (e) Pr = 0.0004; (f-j) Spatial distribution map of copolymer in micelles with

different chain length, corresponding to the upside micelles, the long and short chain are shown in red and blue, respectively.

The SA of P4VP-*b*-PS in methanol, which is the self-assembly progress without the polymerization was also studied. The amphiphilic copolymer system was modelled and their molecular weight and molecular weight distribution were set as Figure 2 (b). Long nanotube was obtained from sphere-vesicle-tube transition as shown in Figure 3 (f). Firstly, the spherical micelles were formed by aggregation of dispersive amphiphilic copolymers due to the solvophobic effect of PS. Then the spherical micelles fused to lamellar structure, which curled to small vesicles with similar size. Finally, the vesicles fused along one direction to form a long nanotube.

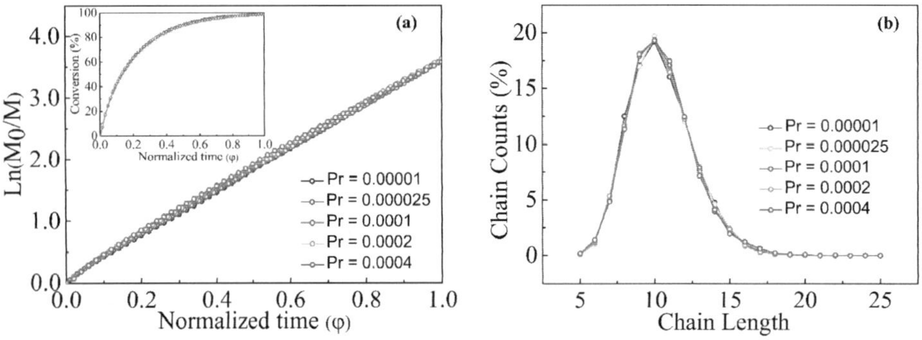

Figure 2. (a) The profile of ln(M_0/M) over time, M_0 is the initial monomer concentration, M is the monomer concentration at a certain time; the inset shows the conversion of the monomers with the normalized time. (b) The probability distribution of copolymer with different chain length in micelles obtained at different Pr.

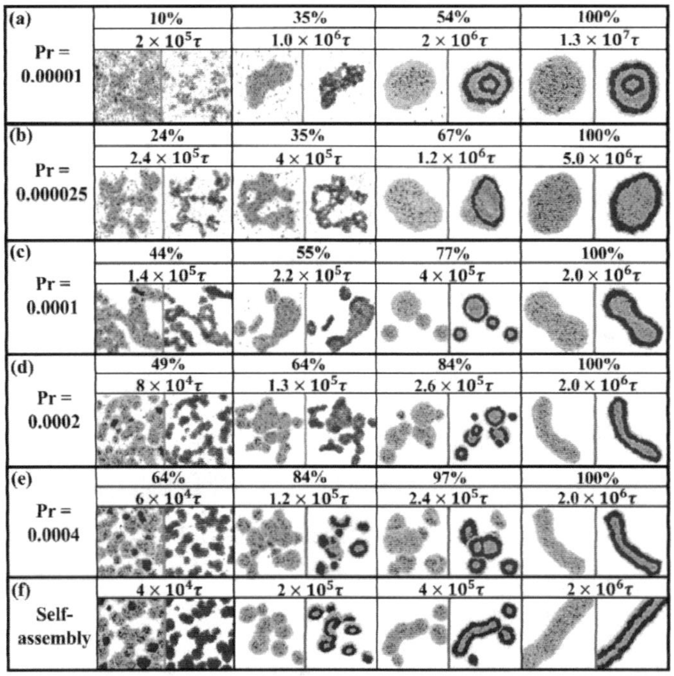

Figure 3. Self-assembly progress of the PISA of five Pr and the SA versus time and the corresponding conversion. Hiding the hydrophilic block and sectional view were used to show the details.

For the PISA, when Pr = 0.0004, the 63.6% of monomers polymerized at $6.0 \times 10^4\tau$ and spherical micelles appeared in Figure 3 (e). Then similar sphere-vesicle-tube transition occurred, and a short nanotube was obtained at 100% conversation. Thus, polymerization at the high Pr didn't affect self-assembly process, because it arrived almost 100% conversion in a short time ($2.4 \times 10^5\tau$). When Pr = 0.0002, similar nanotube was obtained. It is worthy to note that worm-like micelles turned out during sphere-vesicle transition. However, the worms existed in seconds and they transformed into lamella quickly. When Pr = 0.0001, the chain growth rate was further slowed down, so the worms existed for longer time to grow longer and partly crosslink. Then crosslinked worm transformed to a large vesicle, and other short worms curled into small vesicles. Finally, these vesicles with different sizes fused into a peanut-like vesicle because of the different curvature. When Pr decreased to 0.000025, worms crosslinked into a 2D porous structure and then transformed into a large vesicle. When Pr = 0.00001, low polymerization rate led to long time for self-assembly, the worms had enough time to fully crosslink and a porous clew-like sphere was formed. As the chain growth, the holes were filled and the layer was formed from the outer to the inner, finally a multilamellar vesicle was appeared. Therefore, at high Pr, polymerization finished in seconds and weakly affected the self-assembly, that is why PISA behavior at Pr = 0.0004 and the SA is similar. With Pr decreasing, chain growth was slowed down, and transition between spheres to vesicle became slower. The worm-like micelle, which is intervening morphology between spheres to vesicle, turned out. At Pr = 0.0002, worms was transitory and nanotube was formed. With further decrease of Pr, worms existed enough time to grow longer and crosslink, and vesicles with varying architecture were observed. Therefore, changing Pr controlled the existing time of worm-like micelle, which is the key to tune the final morphology to be vesicle or nanotube.

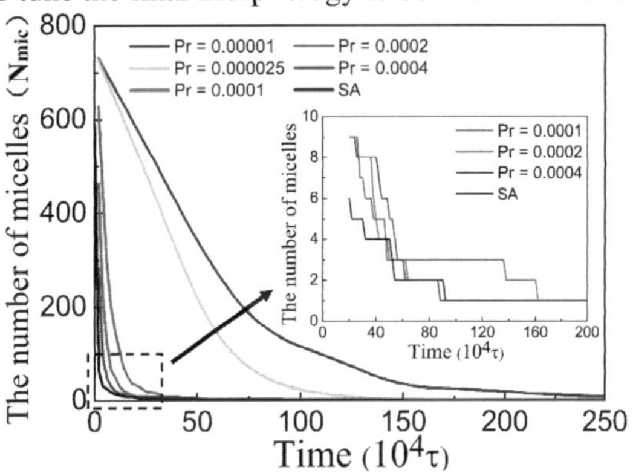

Figure 4. The number of micelles versus time from the PISA of five different Pr and the SA. Inset: An enlarged view of the bottom-left corner.

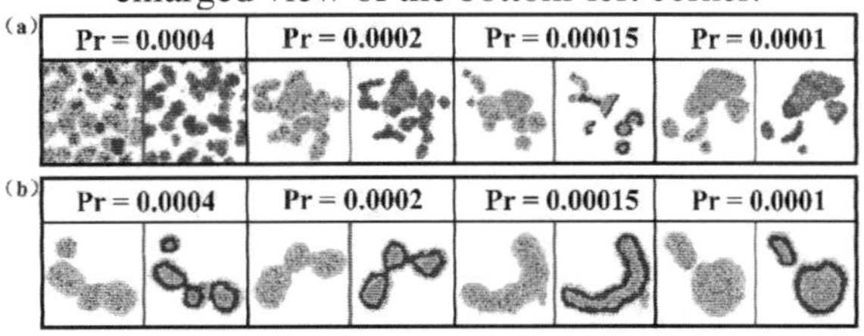

Figure 5. (a) Morphologies when the polymerization was terminated at 64% conversion and (b) final morphologies after continuous evolution.

To further uncover the process of PISA, the number of micelles (Nmic) varies over time were studied as shown in Figure 4. Nmic declined over time both in PISA and SA, due to the fusion of self-assemblies and morphology transition. In PISA, when Pr was increasing, fast growing of hydrophobic block caused Nmic to decrease more quickly, and the Nmic curves of PISA was increasingly close to that of SA. Based on self-assembly process of PISA and SA, we could draw a conclusion that the SA in selective solvent could be regarded as a particular case of the PSIA when Pr is infinite.

In fact, the high conversion is hard to be realized in practical experiments of PISA, especially in organic solvent[3, 10]. The low conversion led to the formation of shorter chain and continuous transformation of morphologies by self-assembly after chain termination. So we also simulated the PISA progress in real experimental condition by terminating the polymerization at ~60% conversion. As shown in Figure 5, at Pr = 0.0004, and 0.0002, vesicles were formed, but these vesicles can't further transfer into tubes with expanding time. These results are well agreed with the vesicles prepared in experiments from PISA of the P4VP-b-PS when the conversion is about 60% to 70%23. Therefore, the low conversion hindered the further transformation of morphology from vesicle to tube. In addition, we adjusted Pr to 0,00015 and the nanotube were successfully synthesized, which showed that the appropriate Pr was the key point to obtain nanotubes even when the conversion was low. However, the region of Pr where nanotubes appeared was relatively narrow, we assumed this is the reason why it is rarely reported about nanotubes obtained in experiments.

3. Conclusion

In summary, PISA of P4VP-b-PS were studied, various morphologies, multilamellar vesicle, peanut-like vesicle, and nanotube, were obtained by tuning Pr. It was found that Pr affected the final morphology no matter the conversation is high (100%) or low (64%). During the PISA, existing time of worm-like micelle as a key point to determine the final morphology is vesicle or tube. Through comparison with PISA, based on the morphologies and their evolution, SA in selective solvent could be regarded as a special case of PISA where Pr is infinite. The application of DPD simulations in PISA strengthens the understanding of PISA and accelerates its further development.

References

[1] (a)Luk, B. T.; Zhang, L. ACS Appl. Mater. Inter. (2014), 6, 21859-21873;
(b)Cash, J. J.; Kubo, T. Bapat, A. P.; Sumerlin, B. S. Macromolecules (2015), 48, 2098-2106;
(c) Shi, P.; Gao, C.; He, X.; Sun, P.; Zhang, W. Macromolecules (2015), 48, 1380-1389.

[2] (a)Semsarilar, M.; Jones, E. R.; Blanazs, A.; Armes, S. P. Adv. Mater (2012), 24, 3378-3382; (b)Jones, E. R.; Semsarilar, M.; Blanazs, A.; Armes, S. P. Macromolecules (2012), 45, 5091-5098; (c)Gao, C.; Zhou, H.; Qu, Y.; Wang, W.; Khan, H.; Zhang, W. Macromolecules (2016), 49, 3789-3798; (d)Zhang, X. Y.; Liu, D. M.; Lv, X. H.; Sun, M.; Sun, X. L.; Wan, W. M. Macromol. Rapid Commun. (2016), 37, 1735-1741.

[3] Wan, W.-M.; Sun, X.-L.; Pan, C.-Y. Macromolecules (2009), 42, 4950-4952.

[4] Warren, N. J.; Mykhaylyk, O. O.; Mahmood, D.; Ryan, A. J.; Armes, S. P. J. Am. Chem. Soc. (2014), 136, 1023-1033.

[5] Warren, N. J.; Armes, S. P. J. Am. Chem. Soc. (2014), 136, 10174-10185.

[6] Gao, C.; Wu, J.; Zhou, H.; Qu, Y.; Li, B.; Zhang, W. Macromolecules (2016), 49, 4490-4500.

[7] Wang, X.; Figg, C. A.; Lv, X.; Yang, Y.; Sumerlin, B. S.; An, Z. ACS Macro Lett., (2017), 6, 337-342.

[8] Sun, X.-L.; Liu, D.-M.; Pei, S.; Li, K.-K.; Wan, W.-M. ACS Macro Lett. (2016), 5, 1180-1184.

[9] (a)Gavrilov, A. A.; Chertovich, A. V. Macromolecules (2017), 50, 4677-4685;
(b)Chen, P.; Yang, Y.; Dong, B.; Huang, Z.; Zhu, G.; Cao, Y.; Yan, L.-T. Macromolecules (2017), 50, 2078-2091.

[10] Zhang, X.; Rieger, J.; Charleux, B. Polym. Chem. (2012), 3, 1502-1509.

IOP Conf. Series: Materials Science and Engineering **859** (2020) 012008 doi:10.1088/1757-899X/859/1/012008

Effect of Wedge Tip Thickness of Nail Clipper on Cutting Characteristics of Polystyrene Ribbon

Thepwachara Ruchirabha, Shigeru Nagasawa

Department of mechanical engineering, Nagaoka University of Technology, 1603-1 Kamitomioka, Nagaoka, Niigata 940-2188, Japan

snaga@mech.nagaokaut.ac.jp

Abstract. This research was aiming to investigate cutting characteristics of Polystyrene ribbon that were cut by nail clippers. A pair of rotational-linked fixture of recent designed JIG was developed for measuring the cutting load and displacement of nail clipper arm. Nail clippers had the upper apex angle of $\alpha_U = 13°$, $\alpha_U' = 45°$ and lower apex angle of α_L=20° and the lower biting edge thickness w_L was chosen 0.06 and 0.25 mm. A 2-mm-width and 1-mm-thickness polystyrene ribbon was cut using the nail-clipper and the load response and the cut trace of sheared zone were investigated under the indentation velocity at $V = 0.01$ mm·s^{-1}. A large force dropped and unstable crack was detected when using $w_L = 0.06$ mm, while the force drop was disappeared without any large crack when w_L= 0.25 mm due to the frictional restriction for fastening the work piece. While the peak line force was direct variation with tip thickness. When considered to vibration during cutting process by using AE sensor, nail clipper that have lower anvil thickness at 0.25 mm have lowest amplitude. Moreover, the cut traces were detected by microscope CCD camera.

1. Introduction

At the presents, nail clipper is almost fundamental equipment for every household. The most popular nail clipper type use nowadays is compound lever type. Typically, ordinary compound lever nail clipper consists of at least three separate rigid parts. Actuator arm that transmits finger force to the jaw and a pin acts as a pivot. Sometimes, there is a file, which assembles to file nail [1].

Many makers of nail clipper try to manufacture a new model of nail clipper blade that can cut a natural nail off with a low resistant load and a smooth sheared profile.

However, such the load response that be gotten from any prototype loading system was insufficient to check for consistency and openness. Therefore, appropriate measurement system of cutting load of developed nail clipper should be verified through a specified condition based experiment.

In the past, Ruchirabha et al. have designed a rotational-linked fixture JIG for using with nipper handle. Through this development of fixture JIG, it was clarified that the capability of designed fixture JIG was consistent for measuring and comparing the cutting response of several nippers [2]. Regarding the relationship between the cutting load (or the cutting moment) and the displacement at the pushing point of nail clipper actuator arm, appropriate theoretical estimation of cutting characteristics of the nail clipper is necessary for understanding the effects of mechanical parameters such as the apex angle of blade, the tip thickness of upper/lower blade.

Content from this work may be used under the terms of the Creative Commons Attribution 3.0 licence. Any further distribution of this work must maintain attribution to the author(s) and the title of the work, journal citation and DOI.
Published under licence by IOP Publishing Ltd

IOP Conf. Series: Materials Science and Engineering **859** (2020) 012008 doi:10.1088/1757-899X/859/1/012008

As for the wedge cutting resistance, since there are several fundamental theories [3-5], the cutting resistance is characterized with the apex angle of wedge and the indentation depth against the thickness of worksheet. One of primary mechanical factors that characterize the cutting resistance is the shear stress. Canakci et.al reported that the shear stress increased with the compressed density of modified expanded polystyrene [6]. Nagasawa et al. reported about the effects of two-line wedge blade on the cutting characteristics of several resin materials [7-9]. The cutting line force and deformation of specimens were affected by the geometry of the blade edge such as the tip radius, the primary height of wedge and the apex angle of blade [10,11].

In a wedge cutting, unstable (multiple deformation modes) shear edge trace (sheared profile) of resin specimen seems to be observed. The multiple deformation modes of shear edge trace were appeared to be caused by the frictional instability or adhesion of wedge surface when choosing a Polycarbonate (PC) and a Polyethylene terephthalate (PET) sheet [12-15]. That provides unstable cracking, bent up necking, large inclination and burred of lower zone.

Baden [16] measured the natural nail stiffness using the speed of sound through nail, the result found that the Young's modulus was approximate twice as high (4.3± 0.3· GPa) transversely as longitudinally (2.1± 0.3· GPa).

Therefore, in this research, in order to reveal the effect of biting edge thickness of nail clipper on the cutting characteristics of a PS ribbon material, the rotational-linked fixture JIG system that developed from [2] was provided for the use of nail clippers that had the represent upper apex angle α_U= 13° with the upper tip thickness w_U= 0.1mm and lower apex angle α_L= 20° with the lower biting edge thickness w_L was chosen as 0.06mm and 0.25mm

2. Materials and experiments

2.1. Specifications of fixture JIG of nail clipper actuator arm and PS specimens

The holding JIG that used in this research were the same as JIG from [4]. It was used newly for the nail clipper. By the actuator arm (handle) of the nail clippers were inserted into the upper socket and arms that connected with jaw were inserted into lower socket for clamping respectively, as shown in Figure 1. The attitude of the blade was arranged to be vertical against the PS ribbon. Because the ribbon specimen was set up across to the nail clipper blade in the horizontal attitude.

A width 2 mm and 1 mm thick of polystyrene (PS) ribbons that cut from TAMIYA B4 size plates item 70124-720, which had a longitudinal length of L=40 mm, was used for the cutting test. All the specimens of PS ribbons were sufficiently cleaned before cutting. The tensile stress-strain relationship in machine direction (MD) (longitudinal direction of bar) of the polystyrene ribbons was examined with the feed velocity $V = 0.01\ \mathrm{mm \cdot s^{-1}}$. In case of mechanical properties of PS, the dimension of polystyrene specimen that used for found stress-strain relationship refer from JIS K7127. The mechanical properties of the PS specimen were shown in Table 1.

Table 1. In-plane tensile mechanical properties of polystyrene (PS) in longitudinal direction (the strain rate: 0.000125 s^{-1} with the span of 80 mm).

Symbol	Thickness t/mm	Young's modulus E/GPa	Yield strength σ_Y/MPa	Tensile strength σ_B/MPa	Breaking strain ε_B
Tensile mode	1	2.74 (2.71–2.79)	16.67 (16.33–16.99)	21.2 (20.42–21.65)	0.21 (0.18–0.25)

Average (Maximum-Minimum measured)

Table 2. Out of plane compressive mechanical properties of polystyrene (PS) in the cutting direction (the y direction). The strain rate: 0.33 s−1.

Symbol	Thickness t/mm	Young's modulus E_C/GPa	Yield strength σ_{YC}/MPa
Compressive mode	3.0	5.04	71.04

After assembling a nail clipper in the fixture JIG on the cutting test apparatus the applied pushing force at the actuator arm F_1 was measured by the load cell when cutting PS ribbon. The cutting force at the blade biting position F_2 was calculated by Eq. 1 and f_2 (= the ratio of F_2 by the width of 2 mm) N/mm. The compressive test of 3 mm square polystyrene bars was measured for $\boldsymbol{\varepsilon} \approx 0.65$ with $V = 1$ mm·s^{-1}. Here, the $\boldsymbol{\varepsilon}$ is the true strain in TD. Figure 3 shows a specimen and the setup of apparatus for compressive test. Through this compressive test, the results were that the compressive Young's modulus E_C in the y direction (cutting direction) and the yield strength σ_{YC} in the y direction (cutting direction) were shown in Table 2. Here, it was noted that this compressive σ_{YC} included a sort of frictional restriction.

From the table 1, found that the young's modulus of PS bar was estimated nearly the young's modulus from the natural nail[16]. Due to the natural nail was hardly using in an experiment. So, the PS bar was used in experiment instead of natural nail. f_2 was the cutting line force. Putting the indentation displacement of the actuator arm D_1, that of the blade biting position D_2 is principally calculated by Eq. 2. Here, since the full stroke value of D_2 is equal to the thickness of specimen, namely D_2 (full stroke) ≈ 1mm in this work. When F_1 was measured during the cutting test, D_1 was measured as the upper crosshead displacement as shown in Figure 1 and the arm span length L_1 and L_2 were 42.91 mm and 5.62 mm on both nail clippers, respectively.

$$F_2 = F_1' \cdot L_1 / L_2 \quad (1)$$

$$D_2 = D_1' \cdot L_2 / L_1 \quad (2)$$

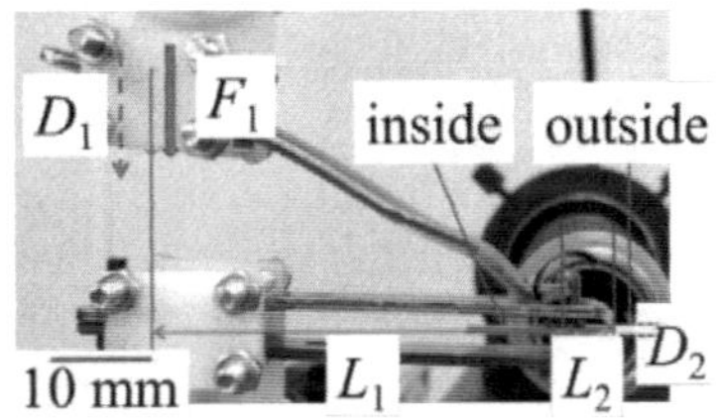

Figure 1. Zoomed up view of a nail clipper and each measured arm span length.

The effective pushing force at the clamped-top position F_1' (N), that excludes the inside-stiffness force of nail clipper from the raw force F_1 measured by the load cell, and is estimated using an idle cut calibration test without any work material. Similarly, the effective upper crosshead displacement D_1', that excludes the elastic deformation of fixture JIG, is calibrated from the measured raw displacement D_1 using the idle cut calibration test without any work material.

The experiments were carried out under the following conditions: a room temperature of 297 K, and a room humidity of 50%RH.

2.2. Specifications of nail clippers and these profile parameters on cutting process

The cutting experiment of a width 2 mm and 1 mm thick of polystyrene (PS) ribbon was carried out using with nail clippers. Figure 2 shows a nail clipper blade which has the upper apex angle $\alpha_U = 13°$ and the back angle (α_U') = 45°. While the lower apex angle (α_L) = 20°. In Figure 3, denote "inside" is represent for the side that natural nail is cut off from the finger and "outside" is represent for the side that remains natural nail which attach with the finger.

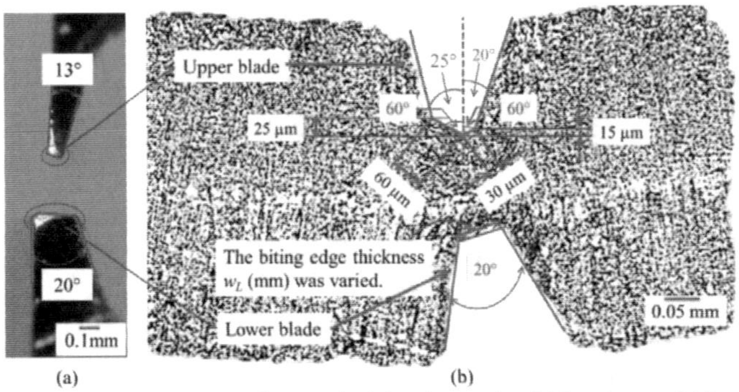

Figure 2. Setup of nail clipper. (a) General side view, (b) Side view of blade from replica.

2.3. Experimental procedure for vibration measurement

In order to know the mechanical vibration during the cutting process, the acoustic emission (AE) system was used for measure the amplitude that occurred. AE transducers were set up at two points: (A) attached with the upper surface of the actuator arm and (B) attached reversed side of the jaw directly.

3. Result and discussions

3.1. Calibration of cutting force with device stiffness at nail clipper's actuator arm.

Regarding the cutting load diagram using the nail clipper, the indentation velocity of blade $V = \mathrm{d}D_1/\mathrm{d}t$ was chosen as 0.01 mm·s^{-1}.

After calibration, it was found that the peak maximum position of raw data was about 49% while that of calibrated data was about 42% of the full stroke. The pushed line force f_1'(= the ratio of F_1' by the width of 2 mm) was monotony increased with D_1' for the first half (less than 25%). This tendency was common for every type of nail clippers. While, in the latter half (larger than 30%), the nail clippers that has the lower biting edge thickness w_L at 0.06mm had a saturated load response and had a breaking point in at $D_1' \approx 92\%$ of 6.5 mm. In the case of the nail clippers that have lower anvil thickness at 0.25mm, it had a stably reducing load response in the latter half (larger than 24% of 8 mm). Then at $D_1' > 87.5\%$ of 8 mm, the upper and lower blade collided each other but due to work hardening that occur at PS ribbon, the specimen didn't break down until the D_1' reach $\approx 97.5\%$ of 8 mm, the PS ribbon break down.

3.2. Cutting response at biting position

The representative cutting stages were classified in 5 stages, as shown in Figure 3 and 4. The first of cutting stages ($D_2 \approx 18\%$ of the height of specimen) was classified as upper/lower blade penetrated into specimen. Then, only the upper blade penetrated into specimen at $D_2 \approx 18\%$ - 76% of the height of specimen and the peak maximum point ocuured at near $D_2 \approx 35\%$ of the height of specimen. Moreover, during the process found that the upper blade penetrated with misalignment of lower anvil as shown in Figure 9 at $D_2 \approx 50\%$ - 78% of the height of specimen. Finally, pre- crack was generated $D_2 \approx 78\%$ of the height of specimen and crack propagated into the uncut area. Then, the specimen was separated.

In case of the nail clipper that has the lower biting edge thickness w_L at 0.25 mm, the represent cutting stages were classified in 5 stages too by only upper blade penetrated until $D_2 \approx 23\%$, as shown in Figure 9. After that during the upper blade penetrated into specimen, the compressive phenomena of lower blade occurred 4 times at 43%, 62% and 83%. After that, any a sort of crack of the specimen didn't occurred, but received work hardening at $D_2 \approx 92\%$. So, the biting line force f_2 increased rapidly. Then, the specimen was cut off at $D_2 \approx 100\%$ under having the biting collision of upper/lower blades.These figures captured by the CCD camera.

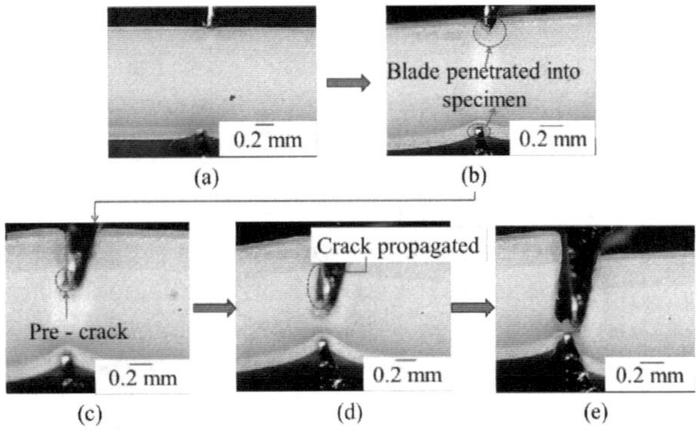

Figure 3. Deformation states of PS ribbon specimen indented by nail clipper that the lower biting edge thickness w_L at 0.06 mm. (a) Machine press actuator arm at D_2 = 0.0 mm, (b) Upper blade/lower anvil penetrated into specimen at $D_2 \approx 0.1$ mm, (c) Pre-crack occur at $D_2 \approx 0.8$ mm, (d) Crack propagated at $D_2 \approx 0.82$ mm, and (e) Cut off at $D_2 \approx 0.87$ mm.

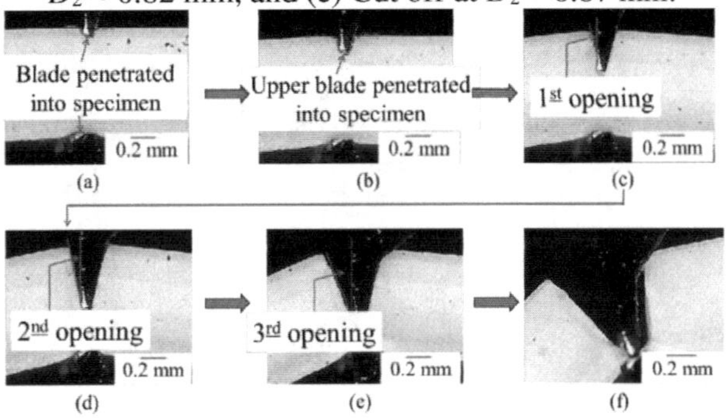

Figure 4. Deformation states of PS ribbon specimen indented by former nail clipper that the lower biting edge thickness w_L at 0.25mm. (a) Upper blade/lower anvil penetrated into specimen at $D_2 \approx$ 0.01 mm, (b)1st lower anvil compress at $D_2 \approx 0.15$ mm, (c) 2nd lower anvil compress at $D_2 \approx 0.40$ mm, (d) 3rd lower anvil compress at $D_2 \approx 0.63$ mm, (e) 4rd lower anvil compress at $D_2 \approx 0.85$ mm and (e) Cut off at $D_2 \approx 1$ mm.

After calibrating the cutting force with device stiffness, Eq 1 used to converted the effective pushing response f_1' (N· mm^{-1}) to blade biting force f_2 (N· mm^{-1}) and Eq 2 were used to converted the indentation displacement of nail clipper's actuator arm D_1' (mm) to the blade biting position D_2 (mm). The relationship between the blade biting line force f_2 and the blade displacement at the biting position D_2 was shown in Figure 5 by each stages refer from Figure 3 and 4.

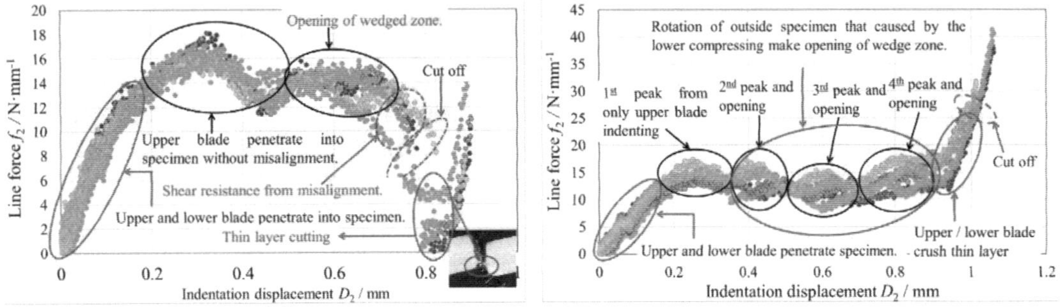

(a) (b)

Figure 5. The blade biting line force f_2 and blade displacement of nail clipper arm D_2 under the indentation velocity V=0.01 mm·s^{-1} of the nail clipper that has the lower biting edge thickness w_L at 0.25 mm.

3.3. Effects of blade structure on sheared profile

After cutting PS ribbon specimens at $V = 0.01\text{mm}\cdot s^{-1}$, the sheared profile of the specimens was observed by a microscope CCD camera. Figures 6(a) and (b) shows the representative side views of sheared zone of PS ribbon specimen. Here, the left of specimen was located in the side that noted as "inside" of Figures 1. The profile parameters of the left sheared edge were defined as h_1(mm), h_2(mm), h_3(mm), and θ (°). That of the right sheared edge were defined as t_1(mm), t_2(mm), t_3(mm), t_4(mm) and w(mm). The parameters of sheared edge profile were measured from the photographs of a microscope CCD camera. Seeing Figure 6(a) and (b), the blade profile was asymmetric with the thickness direction. The existence of the torn zone t_3 seemed to correspond to the stages of propagation of crack.

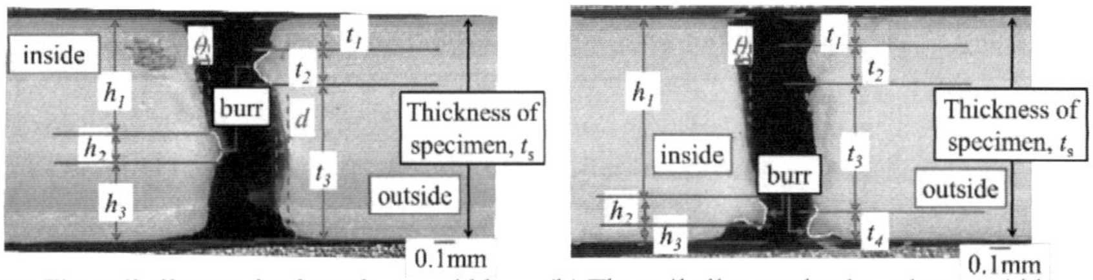

(a) The nail clippers that have lower width at 0.06 mm. (b) The nail clippers that have lower width at 0.25 mm.

Figure 6. Representative sheared profile of PS ribbon specimen that was cut off by using nail clipper that the lower biting edge thickness w_L at (a) 0.06 mm and (b) 0.25 mm.

When choosing the former nail clippers that has the lower biting edge thickness w_L at 0.06 mm, the result show that the pre-crack was detected at $D_2 \approx 0.7$ mm due to low friction force. So, the sheared edge trace has the depth of "*d*" without t_4 zone. In case of the nail clipper that has the lower biting edge thickness w_L at 0.25 mm, any large pre-crack and dispersion of force drop was not detected during the full stroke as seen from Figure 6. Therefore, the nail clipper that have lower anvil thickness at 0.25 mm seems to have a friction force from the lower blade tip for restricted the crack propagation. This friction restriction of lower blade made the shear profile of specimen smoother than another nail clipper as seen in Figure 7(b) (Namely, they don't show any "*d*").

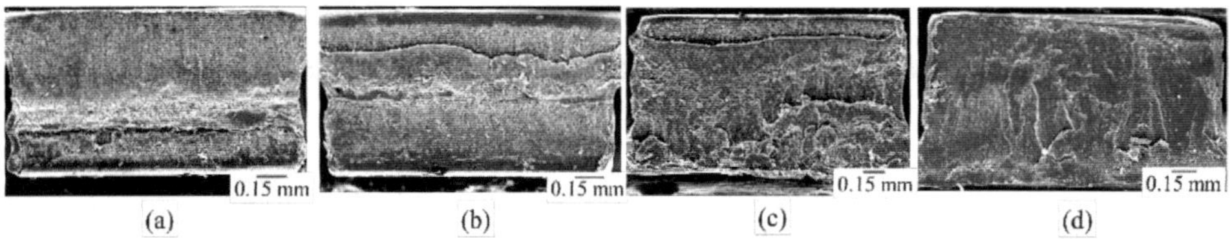

Figure 7. SEM micrographs of sheared surfaces of PS ribbon specimen cut by nail clipper by (a) and (b) are front view of sheared specimen that have cut by lower biting edge thickness wL at 0.06 mm, (c) and (d) are front view of sheared specimen that have cut by lower biting edge thickness wL at 0.25 mm.

The front view of sheared surfaces of specimens cut by the two kinds of nippers were captured by a Scanning Electron Microscope (SEM) and shown in Figure 7(a)-(d). By, the figures 8(a) and (c) were the "outside" and figures 7(b) and (d) were the "inside" of PS ribbon specimen that denoted in Figure 1.

3.4. Effects of blade structure on vibration of nail cutter during cutting process

AE signal could be detected at each set point of the AE transducer. At the setting place of sensors at the actuator arm and jaw, the following AE signals were observed as shown in Figure 8(a) and (b). The result shown that when use the lower blade that has the lower biting edge thickness w_L at 0.06 mm. The peak amplitude was shown 2 position on both sensors. The first peak was detected at the position of pre – crack occur and the second position was detected at the position that specimen was breaking down. While, the cutting process that has the lower biting edge thickness w_L at 0.25 mm, the peak amplitude was detected only one peak amplitude, at the start position of only upper blade indented, as shown in Figure 8. By the peak amplitude was detected at only the jaw due to the indented of upper blade.

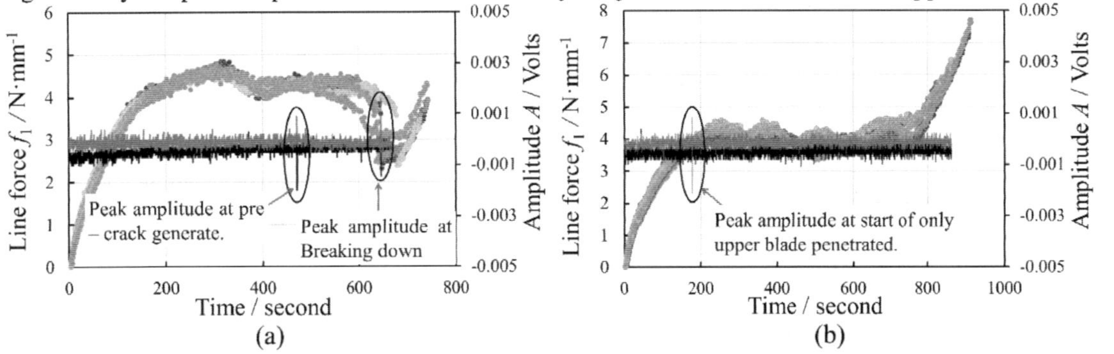

Figure 8. Results of detected AE signals on cutting process when (a) use nail clipper that has the lower biting edge thickness wL at (a) 0.06 mm and (b) 0.25 mm. By the black line is AE signal from sensor B and red line was from sensor A.

From Figure 8(a), it was concluded that the pre-crack generated and breaking down of specimen could make the peak amplitude occurred. Moreover, the open of wedged zone was vibrated in high level for generated the peak amplitude also. When considered to the split off specimen, the less vibration (less maximum amplitude) that make specimen split well could be obtained of the nail clipper that have width of lower blade at 0.25 mm due to the peak amplitude didn't show at the breaking down stage.

4. Conclusion

In order to reveal the effect of blade profile on the cutting line force and the sheared edge profile of a 2mm wide and 1mm thick of polystyrene ribbon, four kinds of nail clippers: the upper/lower asymmetric two-line wedge of angle: $\alpha_U = 13°$, $\alpha_U' = 45°$ against lower side wedge of angle: $\alpha_L = 20°$ that have varied the lower biting edge thickness w_L at 0.06 and 0.25 mm were experimentally investigated. The results were as follow:

(1) In order to perform a certain extent of reproducibility in the cutting test of a nail clipper under a constant indentation velocity, the prototype JIG fixture same as [4] was appropriate when considering the discussed calibration method of idle cutting response.

(2) In case of nail clipper with large lower anvil thickness, the flow stress appeared to be compressive state and a large resistance, while the flow stress appeared to be tensile state and a small resistance value in the case of small lower anvil thickness.

(3) When comparing the sheared edge trace with the nail clippers that varied lower anvil thickness, the nail clipper that have lower anvil thickness at 0.25mm generated a smooth sheared surface without any dynamic large force drop, the pre-crack at the necked stage did not occurred, and also the frictional restriction by the width of tip thickness appeared to fasten the force drop and the dynamic crack propagation.

(4) When varying the tip thickness of anvil, the final force drop is controlled to disappeared for the lower biting edge thickness w_L at = 0.25 mm. It seems to be caused from the frictional fastening.

(5) When considered to the mechanical vibration of specimen that occurred during the cutting process, the wider of the lower blade provided the less mechanical vibration due to frictional restrain.

Acknowledgements
The authors thank Murayama Taiyo of Godhand company for making the nail clippers for experiment.

IOP Conf. Series: Materials Science and Engineering **859** (2020) 012008 doi:10.1088/1757-899X/859/1/012008

References

[1] Shuib S, Ridzwan M, Kadarman A, et al. (2004) The Analysis of a Compliant One-Piece Nail Clipper Using Pseudo-Rigid Body Method in Comparison to Finite Element Method. *Journal of Physical Therapy Science* 15: 69-78.

[2] Ruchirabha T, Nagasawa S (2019) Development of measurement system of cutting load response in nipper blade, and estimation of wedge angle effect on cutting characteristics and shear edge trace of polystyrene bars. *AIMS material Science* 6: 119-144.

[3] Grunzweig J, Longman IM, Petch NJ (1954) Calculations and Measurements on Wedge-Indentation. *J Mech Phys Solids* 2: 81–86.

[4] Johnson KL (1985) Contact Mechanics, *Cambridge Univ. Press*, 160–166.

[5] Hill R (1953) On the Mechanics of Cutting Metal Strips with Knife-edged Tools. *J Mech Phys Solids* 1: 265–270.

[6] Canakci H, Alak D, Celik F (2016) Evaluation of Shear Strength Properties of Modified Expanded Polystyrene Aggregate. *Procedia Eng* 161: 606–610.

[7] Nagasawa S, Masaki Y, Fujikura M, et al. (2011) Analysis of Cutting Characteristic of Polycarbonate Sheet Subjected to Wedge Indentation by Knife Edge and Grooved Plate. *Mach Sci Technol* 15: 110–131.

[8] Nagasawa S, Takahashi T, Fukuzawa Y (2010) Effect of Tool Condition on Wedging Characteristics of Stacked Polycarbonate Sheets. *JP J Solids Struct* 4: 65–84.

[9] Nagasawa S, Masaki Y, Fukuzawa Y, et al. (2012) Bending Effect on Cutting Characteristics of Polycarbonate Thick Sheet Subjected to Indentation of Heightwise-Asymmetric Wedge Blades. *Adv Mater Res* 445: 68–72.

[10] Mitsomwang P, Nagasawa S, Kuroiwa H, et al. (2014) Deformation Analysis of Silicone Rubber Sheet to Keen WC Blade Indentation. *Int J Automation Technol* 8: 761–772.

[11] Mitsomwang P, Natpukkana P, Borrisutthekul R, et al. (2016) Effect of blade tip geometry on cutting characteristics of lead alloy sheet subjected to wedge indentation. *Key Engineering Materials* 719: 137-141.

[12] Kaneko S, Nagasawa S (2020) Wedge indentation characteristics of pressure-sensitive adhesives sandwiched by polyethylene terephthalate films. *Journal of Advanced Mechanical Design, Systems, and Manufacturing* 14: 1-18.

[13] Nagasawa S, Takahashi T, Fujikura M (2010) Contact Stability on Cutting Characteristics of Polycarbonate Sheets Subjected to Two-Line Wedge Indentation. *J Solid Mech Mater Eng* 4: 578–589.

[14] Nagasawa S, Fujikura M, Fukuzawa Y (2011) Finite Element Analysis of Cutting Deformation of Stacked Polycarbonate Subjected to Two-Line Wedge Indentation. *J Solid Mech Mater Eng* 5: 721–731.

[15] Kaneko S, Nagasawa S (2018) Effect of wedge blade direction on cutting characteristics of anisotropic polyethylene terephthalate film. *Transactions of the JSME (in Japanese)* 84: 1–18.

[16] Baden, H. P. (1970). The physical properties of nail. *J. Invest. Dermatol.*55.

IOP Conf. Series: Materials Science and Engineering **859** (2020) 012009 doi:10.1088/1757-899X/859/1/012009

The Effect of Micro-Eggshells and Aluminium Hydroxide on Tracking Resistance of Silicone Rubber Composite

Pattarabordee Khaigunha[1,a], Tanakorn Wongwuttanasatian[2,b], Amnart Suksri[2,c]

[1]Department of Electrical Engineering, Khon Kaen University, Khon Kaen, Thailand
[2]Center of Alternative Energy Research and Development, Khon Kaen University, Khon Kaen, Thailand

[a]pattarabordee.kh@kkumail.com, [b]tanwon@kku.ac.th, [c]samnar@kku.ac.th

Abstract. Break down due to surface tracking is one of the critical insulation failures, especially in the polymeric insulation. It has made worse the reliability of the power transmission system. This present work attempts to solve the problem by observing the effect on tracking resistance of adding micro-eggshells and aluminum hydroxide into the silicone rubber. In order to fabricate the composite specimen, red eggshells (63 to 74 μm), aluminum hydroxide, and RTV silicone rubber were procured. The constant voltage test method, according to the IEC-60587 standard, was used to evaluate the tracking resistance; each specimen was applied by 4.5 kV for 1 hour, and the tracking length was declared as the inverse of tracking resistance. Besides, the rule of mixtures was employed to estimate the thermal conductivity. The results show that the tracking resistance of the composite specimens was overall higher as compare to the pure rubber. Moreover, It was found that tracking resistance and thermal conductivity decreased when the eggshells increased. It was described that the rise of thermal accumulation on the insulator surface in the hybrid composite caused by lower thermal conductivity. Therefore, it could be concluded that the mixing of micro-eggshells and aluminum hydroxide should not be used as fillers for suppressing the tracking failures.

1. Introduction

Silicone rubber has been used as a high voltage outdoor insulator. In the transmission line system, with their advantages such as high dielectric breakdown, UV, and pollution resistance. They are mostly employed to be an external layer for suspension, dead-end, and inter-phase insulators. Even if they have numerous advantages, their surface can be destroyed by surface discharge or tracking, which produces much heat, then break their chemical bond and make the system failure. Thus, it is required to enhance their properties to make them more tracking resistance[1].

There are many efforts to study the problem, and most of them filled inorganic fillers such as silicon dioxide, aluminum oxide, and aluminum hydroxide into the rubber. Among the fillers, aluminum hydroxide is the popular filler. The ability of aluminium hydroxide to release water of hydration at high temperatures (about 220 degrees Celsius) is the crucial function to resist chemical bond breaking by take away heat from the rubber surface[2-4]. Also, there was a previous report which stated that alternative filler as eggshells could improve tracking resistance because it contains absolutely calcium carbonate($CaCO_3$)[5]. By the way, using only eggshells was seem unstable owning to its $CaCO_3$ composition was vary with the source of the eggshells. Therefore, this study investigates the effect of eggshells and aluminium hydroxide on tracking resistance of silicone rubber composite.

Content from this work may be used under the terms of the Creative Commons Attribution 3.0 licence. Any further distribution of this work must maintain attribution to the author(s) and the title of the work, journal citation and DOI.
Published under licence by IOP Publishing Ltd

IOP Conf. Series: Materials Science and Engineering **859** (2020) 012009 doi:10.1088/1757-899X/859/1/012009

2. Materials and methods

Test specimen consisted of 3 composites: room temperature vulcanizing silicone rubber (RTV), micro red eggshells, and aluminum hydroxide (AH). The RTV was supplied in two parts, liquid silicone rubber, and catalyst, by Super Silicone & Resin Art. Micron red eggshells, which contain 93 to 97% calcium carbonate[6], were obtained by grinding and sieving red eggshells until a size of them is in the range of 63 to 74 μm. The AH was supplied by Qchemical. The test specimen was initially fabricated by mixed the rubber with a dedicated amount of each composition, as mentioned in table 1. Then, the composite was stirred by hand for 5 mins to avoid agglomeration. After that, the mixture was stirred again for 2 mins; at the same time, the catalyst was carefully added. Finally, the composite was poured into a 50x120x6 mm. mould, and kept in the oven at room temperature for at least 6 hours.

Tracking resistance was obtained by the test, according to IEC 60587[7]. Tracking resistance defined as the length of a carbon path on the surface of material created by deterioration from the heat of arc discharges. The test is composed of 5 components, such as a 4.5 kV 50 Hz transformer, a 33 kΩ 200 W remitting current resistor, a contaminant feeder with a flow rate of 0.6 ml/min, electrodes, and a specimen mounting support, as shown on the schematic diagram in figure 1. The contaminant is a mixture of 0.1 wt% of NH_4Cl, 0.02 wt% of Triton X-100, and distilled water. For the test procedures, a specimen was firstly mounted on the specimen mounting support at an angle of 45 degrees from the horizontal. Then, assembly the electrodes and eight filter papers. After that, the contaminant was fed to filter papers. While filter papers are full of the contaminant, the contaminant smoothly flows to the bottom electrode through the hole of the top electrode. Finally, the 4.5 kV was supplied for 60 mins to the electrodes. When the testing time reaches 60 mins, the test is done, then the specimen was cleaned by distilled water and measured the length of the deterioration area. Three specimens per material type were tested at the same time for repeatability of results.

Thermal conductivity is the ability of a material to conduct heat. In this work, the rule of mixtures equations is used to calculate the thermal conductivity, so the rule is represented by

$$k_c = k_r V_r + k_{f1} V_{f1} + k_{f2} V_{f2} + k_{f3} V_{f3} + \ldots + k_{fn} V_{fn} \quad (1)$$

In these expressions, k and V denote the thermal conductivity and volume fraction, respectively, whereas the subscripts c, r, and fn represent composite, rubber, and filler number n[8]. In order to calculate the thermal conductivity, the parameters in Table 2 are employed and the results are plotted in figure 4.

Table 1. Composition of test specimens.

Specimen type	Composition (%wt)		
	Silicone rubber	Micro-eggshells	Aluminium hydroxide
E0A0	100	0	0
E1A9	90	1	9
E3A7	90	3	7
E5A5	85	5	5

Table 2. Physical properties of materials in use.

Material type	Physical property		
	k (W/m·K)	Density (g/cm^3)	Sources
Silicone rubber	0.27	1.25	[9]
Micro-eggshells ($CaCO_3$)	2.26	2.77	[9]
Aluminium hydroxide	21.34	2.42	[4]

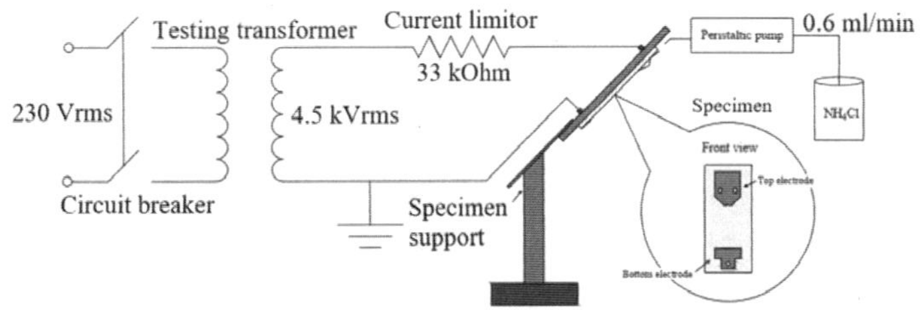

Figure 1. Experimental arrangement for the inclined plane test.

3. Results and discussion

3.1. Tracking resistance

Figure 2 shows the physical development of carbonized tracking patterns on the surface of the test samples with various loadings. The measured physical development of tracking is charted in Figure 3. In general, it can be seen that the pure rubber (E0A0) is the lowest tracking resistance because it has the highest tracking development. Furthermore, it can be observed that tracking resistance is substantially lower in the micro-eggshells/aluminium-hydroxide filled composites with increasing the eggshells contents and decrease the other filler.

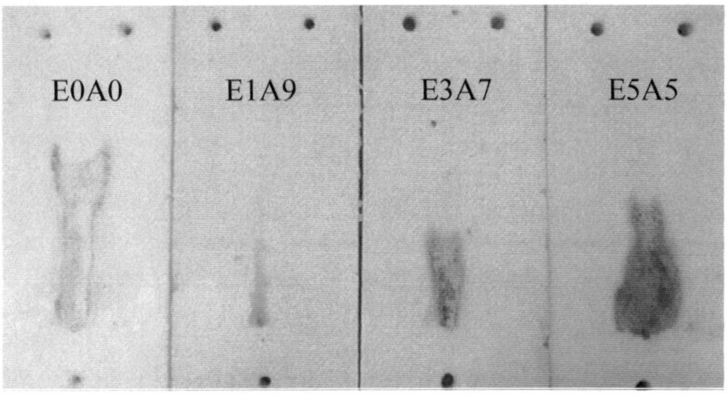

Figure 2. The physical development of tracking patterns after the test.

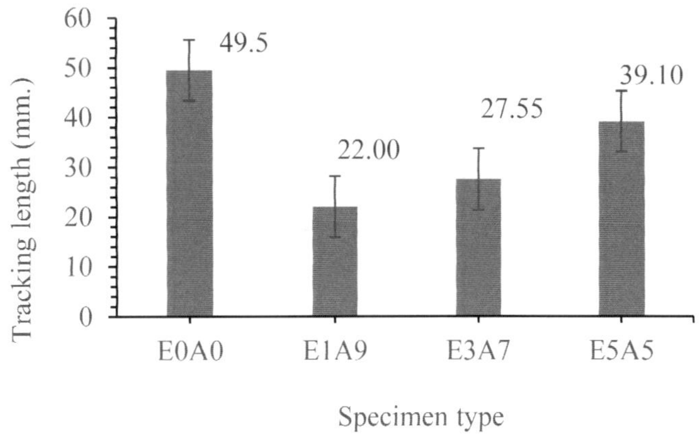

Figure 3. Average tracking length after the test.

3.2. Thermal conductivity

As described early, tracking begins due to temperature build-up on the surface of the composites during the test. Hence, the thermal conductivity of the composites extensively causes the tracking process as it could slow down the arc discharge mechanism and reduce the tracking failure possibilities. As shown in Figure 4, the thermal conductivity of hybrid composites is relatively higher as compared to the A0E0 specimen, but it decreases when the eggshells increase. With decreasing thermal conductivity, hybrid composites can conduct heat away less effectively in the discharge area during the test. Therefore, the tracking failure trends to growth on the composite specimens.

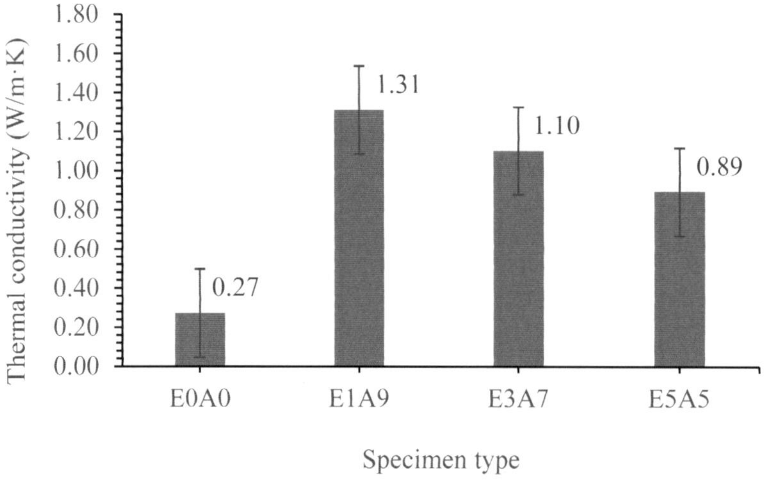

Figure 4. Thermal conductivity of specimens.

In order to explain the decrease in thermal conductivity of hybrid composites, the decomposition of aluminium hydroxide and calcium carbonate is engaged. When the arc discharge occurs, it generates the heat and transfers it to the fillers through the rubber. As a result, the chemical bonds of the fillers tend to break when the temperature reaches their melting point. The melting point of aluminium hydroxide is about 220 C°; at this point, aluminium hydroxide will release water and form aluminium oxide, as shown in equation 2. The released water is considered as the primary function to takes away the heat from the discharge region. Adding both aluminum hydroxide and micro-eggshells shows the result in a counter way. It is considered that calcium carbonate in the eggshells changes into calcium oxide at approximately 800 C° [10] by arc discharge, as shown in equation 3. Calcium oxide absorbs the released water and forms calcium hydroxide, as shown in equation 4. At this stage, the amount of water decrease and thermal conductivity decrease[11].

$$2Al(OH)_3 \rightarrow Al_2O_3 + 3H_2O \quad (2)$$

$$CaCO_3 \rightarrow CaO + CO_2 \quad (3)$$

$$CaO + H_2O \rightarrow Ca(OH)_2 \quad (4)$$

4. Conclusion

In this study, the effect of micro-eggshells and aluminium hydroxide on tracking resistance of silicone rubber composite has been investigated. Due to the water-absorbing of calcium oxide, the composites have lost their thermal conductivity and increase the chance of tracking failure. It could be observed that adding the micro-eggshells and aluminium hydroxide into silicone rubber may lead to lower the tracking resistance. Therefore, it is suggested that the mixing of micro-eggshells and aluminum hydroxide should be used at appropriate amount for the purpose of fillers for suppressing the tracking failures. However, further study on other properties of silicone rubber with the loading of micro-eggshells and aluminium hydroxide is needed.

Acknowledgements

This research study is financially supported by the Center for Alternative Energy Research and Development, Faculty of Engineering, Khon Kaen University, Thailand.

References

[1] A. R. Verma and B. Subba Reddy, "Interpretation of surface degradation on polymeric insulators," Eng. Fail. Anal., vol. 95, pp. 214–225, 2019.

[2] M. Tariq Nazir, B. T. Phung, S. Yu, and S. Li, "Effects of thermal properties on tracking and erosion resistance of micro-ATH/AlN/BN filled silicone rubber composites," IEEE Trans. Dielectr. Electr. Insul., vol. 25, no. 6, pp. 2076–2085, Dec. 2018.

[3] L. H. Meyer, E. A. Cherney, and S. H. Jayaram, "The role of inorganic fillers in silicone rubber for outdoor insulation alumina tri-hydrate or silica," IEEE Electr. Insul. Mag., vol. 20, no. 4, pp. 13–21, 2004.

[4] L. Meyer, V. Grishko, S. Jayaram, E. Cherney, and W. W. Duley, "Thermal characteristics of silicone rubber filled with ATH and silica under laser heating," in Conference on Electrical Insulation and Dielectric Phenomena (CEIDP), Annual Report, 2002, pp. 848–852.

[5] M. Hasnun and A. Hassan, "(Lecture Notes in Mechanical Engineering) Mohd Hasnun Arif Hassan-Intelligent Manufacturing & Mechatronics-Springer Singapore (2018).pdf," no. January. 2018.

[6] M. Kristl, S. Jurak, M. Brus, V. Sem, and J. Kristl, "Evaluation of calcium carbonate in eggshells using thermal analysis," J. Therm. Anal. Calorim., vol. 138, no. 4, pp. 2751–2758, Nov. 2019.

[7] I. E. Commission, "IEC 60587 Electrical insulating materials used under severe ambient conditions–Test methods for evaluating resistance to tracking and erosion." IEC publications, Switzerland, 2007.

[8] W. D. Callister and D. G. Rethwisch, Materials science and engineering: an introduction. Wiley New York, 2018.

[9] "Materials Database - Thermal Properties - Thermtest Inc." [Online]. Available: https://thermtest.com/materials-database. [Accessed: 27-Feb-2020].

[10] R. Mohadi, K. Anggraini, F. Riyanti, and A. Lesbani, "Preparation Calcium Oxide From Chicken Eggshells," Sriwij. J. Environ., vol. 1, pp. 32–35, Aug. 2016.

[11] Y. Koshino, I. Nakajima, and I. Umeda, "Effect on the electrical properties of fillers in silicone rubber for outdoor insulation," in Proceedings of the International Symposium on Electrical Insulating Materials, 1998, pp. 465–468.

IOP Conf. Series: Materials Science and Engineering **859** (2020) 012010 doi:10.1088/1757-899X/859/1/012010

Investigation on Solubility of Carbon Dioxide in the Mixed Aqueous Solution of MEA and 2-MAE

Thidarat Meekoch, Kreangkrai Maneeintr*

Carbon Capture, Storage and Utilization Research Laboratory, Department of Mining and Petroleum Engineering, Faculty of Engineering, Chulalongkorn University, Bangkok 10330, Thailand.

*Krengkrai.M@chula.ac.th

Abstract. Absorption with amine solution is a method to reduce CO_2 causing climate change. The commercial amine solution is MEA. The disadvantage of MEA is low CO_2-loading capacity. Also, 2-(methylamino)-ethanol or 2-MAE is a new solvent developed to improve absorption performance. The aim of this work is to measure the CO_2 solubility in the mixture of MEA and 2-MAE at 15:15 %wt, from 30 °C to 80 °C and CO_2 partial pressures ranging from 5 to 100 kPa. From the results, at the same conditions, 30 %wt of MEA has higher absorption capacity than that of 15:15 %wt of MEA:2-MAE and 30 %wt of 2-MAE for 4.51 % and 7.04 %, respectively. For cyclic capacity, 15:15 %wt of MEA:2-MAE has capacity greater than 30 %wt of MEA for 34.10 % but lower than 30 %wt of 2-MAE for 22.78 %. A mixed-amine solution can be applied to reduce the disadvantage of MEA solution.

1. Introduction

CO_2 is one of the greenhouse gases (GHGs) that cause an increase in global warming [1]. CO_2 is one of the by-products of all combustion processes due to the consumption of fossil fuel energy and especially from the power plants [2]. There are many ways to capture the huge amount of CO_2 before releasing to the atmosphere such as chemical and physical adsorption, absorption, cryogenics separation and membrane separation [3]. Chemical absorption by using amine solution is an alternative method to reduce CO_2 for post-combustion process from power plants [4]. The most widely used amine solution is MEA because of the high efficiency of CO_2 absorption performance and fast absorption rate. However, the disadvantages of MEA are low CO_2 loading capacity, corrosion and high energy consumption for solvent regeneration [5]. Also, 2-MAE is a new solvent that has been developed and applied to improve CO_2 absorption performance. However, the disadvantage of 2-MAE is high cost and higher heat of regeneration than MEA [6]. Thus, these solvents are mixed together to achieve higher efficiency. Therefore, the aim of this work is to measure the solubility and investigate the effect of parameters on the solubility of CO_2 in the mixture of MEA and 2-MAE at concentration of formulated solution, the partial pressures of CO_2 in a range of 5 to 100 kPa and temperature in a range of 30 ° C to 80 ° C. The results are compared with these of MEA and 2-MAE solutions at the same conditions. The rate of CO_2 loading change with time of MEA, 2-MAE and mixed amine solution are also compared.

Content from this work may be used under the terms of the Creative Commons Attribution 3.0 licence. Any further distribution of this work must maintain attribution to the author(s) and the title of the work, journal citation and DOI.
Published under licence by IOP Publishing Ltd

IOP Conf. Series: Materials Science and Engineering **859** (2020) 012010 doi:10.1088/1757-899X/859/1/012010

2. Materials and methods

2.1. Materials

An industrial-grade cylinder of CO_2 and Nitrogen (N_2) are supplied by Linde, Thailand Public Company Limited with the purities of 99.5 % and 99.5 %, respectively. MEA with a purity of > 99.0 % is purchased from Sadara Chemical Company. 2-MAE with a purity of ≥ 98.0 % is purchased from Merck. Hydrochloric acid (HCl) at 1.0 N is purchased from RCI Labscan Limited. All amine solutions are prepared to concentration with de-ionized water from PURELAB Classic DI.

2.2. Experimental procedures

In this study, MEA and 2-MAE are mixed at concentration 15:15 %wt with de-ionized water to prepare an amine solution and fed into a reactor which is placed into the temperature-controlled water bath to maintain condition in a range of 30 ° C to 80 ° C. CO_2 and N_2 are mixed together and controlled by mass flow meters until it reaches the desired partial pressure of CO_2 and N_2 in a range of 5 to 100 kPa. In the reactor, CO_2 is absorbed in the solution and sent to the condenser to recover the moisture in the gas stream. The operation is continued to the equilibrium condition taking approximately 12-20 hours. The samples are taken for 3 times to get the average CO_2 loading at equilibrium. The solubility of CO_2 data is obtained by titration using 1.0 N HCl with methyl orange as the indicator.

3. Results and discussion

3.1. Verification

To ensure the experimental results, the procedure and equipment have been verified by comparing with the results of Shen and Li [7], Jou et al. [5], Aronu et al. [8], Yamada et al. [9] and Luemunkong [10] before running the experiment of the solubility of CO_2. Testing conditions are at 5.0 M of MEA solution, the temperature of 40 ° C and pressures used in the verification are 5 to 100 kPa and it can be seen that the results get along well with the ones from the previous works. The results of the equipment verification are shown in Figure 1. The average absolute deviation (%AAD) [11] compared with literatures [10] and [7] are 1.05 % and 1.06 %, respectively. Consequently, it can be concluded that the procedure and equipment are valid in order to perform a further study of this work.

3.2. Effect of type of solvents

Figure 2 shows the solubility of CO_2 in 30 %wt of 2-MAE, 15:15 %wt of MEA:2-MAE and 30 %wt of MEA at the temperature of 40 ° C and 80 ° C and partial pressure of CO_2 at 15 kPa. The results of CO_2 loading at 40 ° C shows that 30 %wt of MEA provides higher absorption capacity than that of 15:15 %wt of MEA:2-MAE and 30 %wt of 2-MAE for 4.51 % and 7.04 %, respectively. At 80 ° C, it shows that 30 %wt of MEA provides much higher absorption capacity than that of 15:15 %wt of MEA:2-MAE and 30 %wt of 2-MAE for 11.15 % and 19.38 %, respectively. This means that 30 %wt of 2-MAE can absorb less at regeneration condition. At partial pressure lower than 15 kPa, 30 %wt of 2-MAE provides less absorption capacity than the mixed amine solution and 30 %wt of MEA especially at the higher temperatures.

3.3. Effect of temperature and partial pressure

Figure 3 represents the solubility of CO_2 in each amine solution at the partial pressure of CO_2 from 5 to 100 kPa and temperature from 30 ° C to 80 ° C. The results can be observed that the solubility of CO_2 at higher temperatures have less capacity when compared with the solubility of CO_2 at lower temperatures. The absorption process is considered as absorption of CO_2 at low temperature when the temperature increases, it results in the lower CO_2 loading capacity. At high temperature, it is considered as a regeneration condition since the gas is removed from the solvent. In this study, the temperature at 40 ° C and 80 ° C are considered as the absorption and regeneration conditions,

IOP Conf. Series: Materials Science and Engineering **859** (2020) 012010 doi:10.1088/1757-899X/859/1/012010

respectively. Thus, it leads to an explanation of the cyclic capacity. In addition, the partial pressure of CO_2 increases from 5 to 100 kPa will become the mass driving force of CO_2 in the gas phase transfer to the liquid phase. Therefore, the higher of partial pressure of CO_2 can capture the higher amount of CO_2.

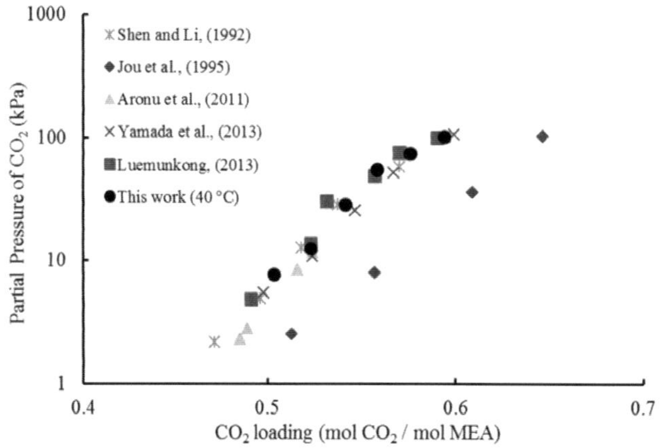

Figure 1. Comparison of the solubility of CO_2 in 5.0 M MEA solution at 40 ° C.

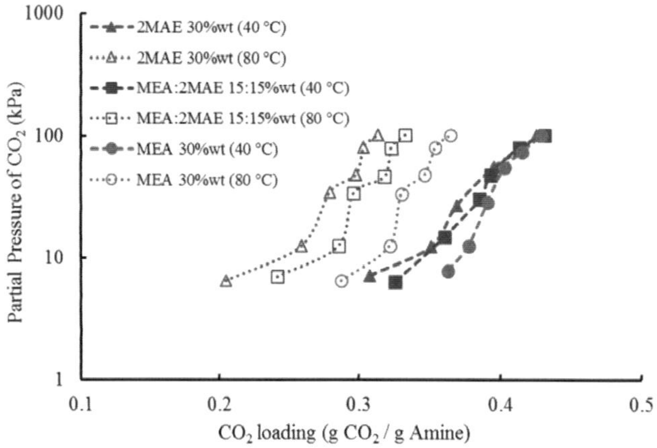

Figure 2. Comparison the solubility of CO_2 in three amine solutions at 40 ° C and 80 ° C.

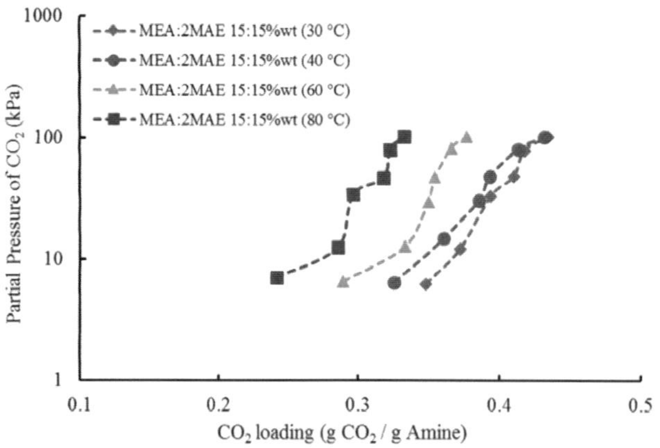

Figure 3. The solubility of CO_2, 15:15 %wt of MEA:2-MAE, from 30 ° C to 80 ° C and 5 to 100 kPa

3.4. Cyclic capacity of the solution

The cyclic capacity is clarified as the difference in CO_2 solubility at the absorption and regeneration conditions [9]. The CO_2 loading is derived from the mole and/or mass of absorbed CO_2 per mole and/or mass of amine. In this study, the temperature 40 ° C of absorption condition and 80 ° C of regeneration condition. Thus, the temperature is important that affects the CO_2 capture capacity of the solvent and the increasing cyclic capacity will reduce the operating cost and CO_2 capture cost [12]. Figure 4 shows the cyclic capacity of 30 %wt of 2-MAE, 15:15 %wt of MEA:2-MAE and 30 %wt of MEA at 15 kPa. The results show that 15:15 %wt of MEA:2-MAE has greater cyclic capacity than that of 30 %wt of MEA for 34.10 % but lower than that of 30 %wt of 2-MAE for 22.78 % at the same condition.

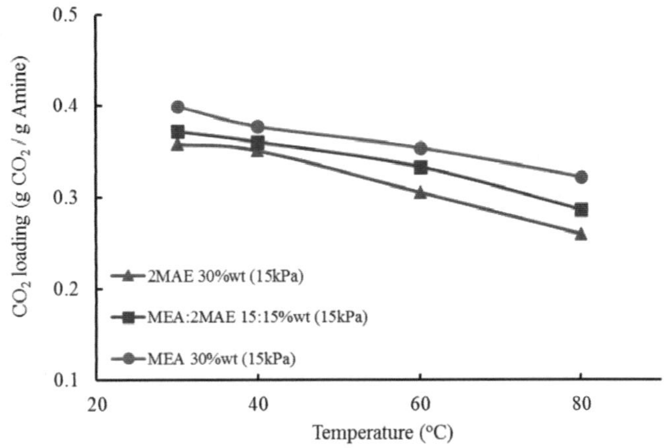

Figure 4. Cyclic capacity of three amine solutions at 15 kPa.

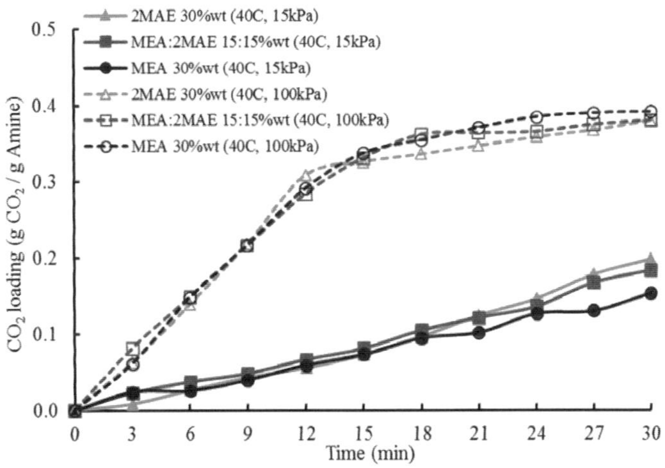

Figure 5. Rate of CO_2 loading change in three amine solutions at 40 °C

3.5. Rates of CO_2 loading change

Figure 5 represents the rate of CO_2 loading change in 30 %wt of 2-MAE, 15:15 %wt of MEA:2-MAE and 30 %wt of MEA at 40 ° C has been studied. In the first 30 minutes at 15 kPa, it can be seen that the rate of CO_2 loading of 30 %wt of 2-MAE can be reacted and higher absorption capacity than that 15:15 %wt of MEA:2-MAE and 30 %wt of MEA because 30 %wt of 2-MAE is required to form unstable carbamate when hydrolysis with water, it becomes the presence of more bicarbonate and less carbamate because one free amine of 2-MAE solution. Therefore, more bicarbonate leads to higher CO_2 absorption capacity. And the rate of CO_2 loading change at 100 kPa, it presents slightly change absorption capacity of three amine solutions because of the partial pressure of CO_2 increases as the solubility of CO_2 increases due to the higher diving force leading to the rate of reaction.

4. Conclusions

The solubility of CO_2 is investigated in mixed amine solution at 15:15 %wt of MEA:2-MAE compared with 30 %wt of MAE and 30 %wt of 2-MAE in the same conditions at the partial pressure of CO_2 from 5 to 100 kPa and temperature ranging from 30 ° C to 80 ° C. The results represent that the solubility of CO_2 increases as the partial pressure of CO_2 increases whereas the CO_2 loading decreases as the temperature increases. The condition at 30 %wt of MEA is the highest CO_2 loading whereas at 30 %wt of 2-MAE is the lowest CO_2 loading on the mass bases. In the effect of temperature, it can be concluded that at 40 ° C is the optimum condition to absorb the CO_2 over a range of 0.360 grams of CO_2 per gram of amine. In term of cyclic capacity, 15:15 %wt of MEA:2-MAE has more cyclic capacity than 30 %wt of MEA for 34.10 % whereas lower than 30 %wt of 2-MAE for 22.78 %. Meanwhile, the rate of CO_2 loading change is becoming higher as the CO_2 partial pressure increases due to an increase in driving force in the liquid phase. Therefore, the CO_2 loading at 100 kPa is higher than CO_2 loading at 15 kPa.

Acknowledgments

The authors would like to gratefully acknowledge the Malaysia-Thailand Joint Authority (MTJA) for financial support of this project.

References

[1] Duro J A 2016 *Ecological Indicators* 66 173
[2] Wang Y, Zhao L, Otto A, Robinius M and Stolten D 2017 *Energy Procedia* 114 650
[3] Leung D Y C, Caramanna G and Valer M M M 2014 *Renewable and Sustainable Energy Reviews* 39 426
[4] Liang Z, Rongwong W, Liu H, Fu K, Gao H, Cao F, Zhang R, Sema T, Henni A, Sumon K, Nath D, Gelowitz D, Srisang W, Saiwan C, Benamor A, Marri M A, Shi H, Supap T, Chan C, Zhou Q, Zahra M A, Wilson M, Olson W, Idem R and Tontowachwuthikul P 2015 *International Journal of Greenhouse Gas Control* 40 26
[5] Jou F Y, Mather A E and Otto F D 1995 *The Canadian journal of chemical engineering* 73 140
[6] Shi H, Zhou Y, Si M, Zuo Y, Kang S, Wang Y, Cui L, Idem R, Tontiwachwuthikul P and Liang Z 2017 *The Canadian journal of chemical engineering* 95 1471
[7] Shen K P and Li M H 1992 *Journal of chemical and Engineering Data* 37 96
[8] Aronu U E, Gondal S, Hessen E T, Warberg T H, Hartono A, Hoff K A and Svendsena H F 2011 *Chemical Engineering Science* 66 6393
[9] Yamada H, Chowdhury F A, Goto K and Higashii T 2013 *International Journal of Greenhouse Gas Control* 17 99
[10] Luemunkong T 2013 *Carbon dioxide Absorption in Aqueous Solution of Dimethylaminoethanol* (Master's Thesis in Chemical Engineering, Department of Chemical Engineering, Faculty of Engineering, Chulalongkorn University)
[11] Mondal B K, Bandyopadhyay S S and Samanta A N 2017 *Chemical Engineering Science* 170 58
[12] Bernhardsen I M and Knuutila H K 2017 *International Journal of Greenhouse Gas Control* 61 27

Measurement of Mass Transfer Coefficient of CO_2–Amine System from Absorption Process

Thanakornkan Limlertchareonwanit[1], Kreangkrai Maneeintr[2,*], Tawatchai Charinpanitkul[1]

[1]Center of Excellence in Particle Technology, Department of Chemical Engineering, Faculty of Engineering, Chulalongkorn University, Bangkok 10330, Thailand.
[2]Carbon Capture, Storage and Utilization Research Group, Department of Mining and Petroleum Engineering, Faculty of Engineering, Chulalongkorn University, Bangkok 10330, Thailand.

*Krengkrai.M@chula.ac.th

Abstract. Carbon dioxide (CO_2) is one of the main factors for climate change issues. Absorption is the technology to remove CO_2 from various industries. The equipment used is the packed column to promote high mass transfer rate and to study the overall mass transfer coefficient (K_Ga_v). The K_Ga_v is an important parameter for designing and scaling up of packed column. Therefore, the objective of this work is to measure K_Ga_v for CO_2 absorption into the aqueous solution of monoethanolamine (MEA) in a random-packed column and to investigate effect of parameters on K_Ga_v. The parameters are the solvent flow rate from 5.3-15.9 m^3/m^2h, CO_2 initial loading from 0.0-0.2 mole/mole and CO_2 content in gas phase from 5-15 %v/v. The results show that K_Ga_v increases as the solvent flow rate and solvent concentration increase. The CO_2 content in gas phase provides relatively high impact on K_Ga_v with an increase in the CO_2 content.

1. Introduction

Currently, the major problem of global warming is an increase in the amount of greenhouse gases (GHGs) in the atmosphere. Carbon dioxide (CO_2) is the main component of GHGs up to 81% [1]. CO_2 is usually produced from three main processes; pre-combustion, post-combustion and oxy-fuel combustion. Coal-fired power plant is the major industry to release the CO_2 to the atmosphere. Thus, there are many researchers interested in carbon dioxide capture and storage (CCS) from post-combustion process in coal-fired power plant. Absorption technology is commonly used absorption solution, like monoethanolamin (MEA). MEA is a primarily chemical solution because of high reactivity, rapid reaction rate, low cost [2], low molecular weight and uncomplicated structure whereas the disadvantages are high energy consumption and corrosion [3]. The packed column is a basic equipment to study the mass transfer coefficient which is important for designing and scaling up of packed absorption column in the future. The features of this column can be operated in the ambient temperature and pressure.

Two Film Theory is suggested by Whitman [4] about the chemical reaction between gases into liquid phase. The liquid is soaked onto the packing with a thin film and the gas is dissolved and reacted with the liquid on the packing. The gas-phase mass transfer coefficient can be calculated in

Content from this work may be used under the terms of the Creative Commons Attribution 3.0 licence. Any further distribution of this work must maintain attribution to the author(s) and the title of the work, journal citation and DOI.
Published under licence by IOP Publishing Ltd

terms of K_G. However, the overall volumetric mass transfer coefficient is more useful based on the unit volume of the packed column. The overall volumetric mass transfer coefficient can be determined using equation (1);

$$K_G a_V = \left(\frac{G_I}{P(y_{CO_{2,g}} - y^*_{CO_{2,i}})}\right)\left(\frac{dY}{dZ}\right) \tag{1}$$

where

$K_G a_v$ is the overall volumetric mass transfer coefficient ($kmol/m^2 \cdot h \cdot kPa$)

a_v is specific area of contactor (m^2/m^3)

G_I is inert gas flow rate ($kmol/m^2 \cdot h$)

P is total pressure (kPa)

$y_{CO_{2,g}}$ is mole fraction of CO_2 in the bulk gas phase

$y^*_{CO_{2,i}}$ is mole fraction of CO_2 in bulk liquid that equilibrium with CO_2 concentration in bulk gas

Maneeintr et al. [5] investigated the mass transfer coefficient of MEA solution in a laboratory-scale absorption column packed with DX structured packing. It is reported in terms of mass transfer to the overall mass transfer coefficient. It is found that inert gas flow rate has an insignificant effect on mass transfer.

Halim et al. [6] studied the mass transfer coefficient of MEA at high-pressure condition (5.0 MPa). The experiment is conducted in the absorption packed column with Sulzer metal gauze packing. The results show that the temperature at 40°C is the optimum condition to remove the CO_2. $K_G a_v$ has increased with an increased solvent flow rate and solvent concentration. At the 5.0 MPa, it can be explained that the performance to remove CO_2 is similar at the atmosphere.

Naami et al. [7] studied of the mass transfer coefficient in a lab-scale packed absorption column with high-efficiency DX structured packing. The experiment has conducted at pressure atmosphere and 14.9 v/v% gas mixture. This study compares the mass transfer between MEA and MDAE. The result show that $K_G a_v$ increase as the liquid flow rate increased. $K_G a_v$ of MEA is higher than MDEA because MEA is a primary amine and high reactivity than tertiary amine.

Therefore, the objective of this study is to measure overall volumetric mass transfer ($K_G a_v$) for CO_2 absorption into the aqueous solution of monoethanolamine (MEA) in a random-packed column and to investigate effect of parameters on $K_G a_v$ and design a random packed column in the future.

2. Experimental

2.1. Chemical

Monoethanolamine (MEA) with purity of 99.0% is purchased from Sadara Chemical Company. Hydrochloric acid (HCl) with purity of 99% is purchased from RCI Labscan Limited. De-ionized water from Purelab Classic is used to prepare a MEA solution. N_2 and CO_2 with the same purity 99.5% are purchased from Praxair, Thailand.

2.2. Experimental apparatus

The absorber column is in 1.8-m high and 25-mm diameter with randomly packed column. The packed column is made of borosilicate glass. Raschig rings (with stainless steel, AISI304) are used as packing for this study. The advantages of stainless steel raschig rings are inexpensive, low pressure drop, free volume more than ceramic, resistance to temperature and physical force and resistance to amine and alcohol compared with ceramic raschig rings [8]. Amine solution is prepared to the desired concentration by mixing deionized water. 1M HCl is used to verify the amount of CO_2 in amine solution after the absorption process in a column by titration with methyl orange as an indicator. The CO_2 loading of each solution is determined by the Association of Official Analytical Chemists [9] and can be calculated from mole CO_2 / mole amine [10].

IOP Conf. Series: Materials Science and Engineering **859** (2020) 012011 doi:10.1088/1757-899X/859/1/012011

2.3. Operating procedure
The flue gases (N_2 and CO_2) is entered to the mass flow controller (Aalborg) and fed to the bottom of the column, while the amine solution is pumped by a peristaltic pump (Longer) and the flow rate is controlled with rotameter (Cole-Palmer) to the top of the column. The gas outlet leaves at the top and the liquid outlet leaves at the bottom of the column. The CO_2 content in gas phase is checked with the IR analyzer within a reliable measurement range of 0-20 v/v%. The operating condition is shown in Table 1.

For this column, there are 11 sampling points along the height of the column. At each point, there are pressure gauges, temperature probes and CO_2 sampling point to measure the concentration of CO_2 in gas phase. The column is filled with raschig rings for increasing contact area between the gas-liquid phases. Flue gas and amine solution contact each other within the column. When the steady state is reached, flue gas is sent to analyzer to measure the amount of CO_2 in the system at each point. After amine solution has already reacted with CO_2, it leaves to the bottom of the column. Flue gas leaves at the top of the column.

Table 1. Operating condition of the experiments.

Variable	Range
Solvent flow rate, $m^3/m^2 \cdot h$	5.3, 10.6, 15.9
CO_2 content in gas phase, v/v%	5, 10, 15
CO_2 initial loading, mole CO_2 / mole amine	0.0, 0.1, 0.2

3. Results and discussion

3.1. Effect of solvent flow rate
The solvent flow rate strongly affects the mass transfer coefficient. The effect of solvent flow rate on K_Ga_v is shown in Figure 1. It can be seen that the solvent flow rate at 15.9 $m^3/m^2 \cdot h$ provides the highest K_Ga_v at 0.37 $kmol/m^3 \cdot h \cdot kPa$. The solvent flow rate is a direct relationship with the K_Ga_v [11], the K_Ga_v increase as the liquid flow rate increase. The reasons for this behaviour are (i) an increase in the solvent flow rate can cause more amine molecules available for reacting with CO_2 molecules, (ii) an increase in the mass transfer in liquid phase which is directly proportional to the K_Ga_v increase, and (iii) an increase in the solvent flow rate leads to increasing the wet surface area between gas and liquid phase on the packing [12].

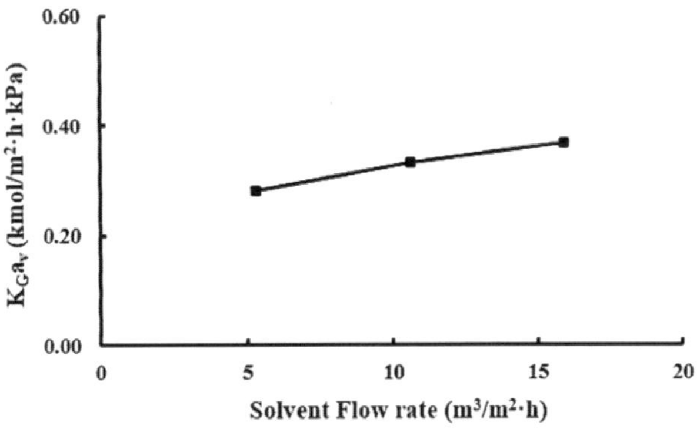

Figure 1. Effect of solvent flow rate on K_Ga_v. Solvent concentration = 3 $kmol/m^3$; solvent flow rate = 5.3-15.9 m^3/m^2h; CO_2 content in gas phase = 15 v/v%; CO_2 loading = 0.0 mole CO_2/mole amine.

3.2. Effect of CO_2 content in gas phase
The effect of CO_2 content in gas phase is the concentration of CO_2 in the gas stream after burning the fossil fuel. According to Figure 2, the CO_2 content at 5 v/v% has the highest K_Ga_v because of the

partial pressure of CO_2 increase leads to K_Ga_v decrease. It can represent to two-film theory, k_L is a main of mass transfer coefficient due to CO_2 absorption is controlled by the liquid phase. An increase in the driving force results in CO_2 partial pressure as K_Ga_v decrease. This result corresponds well with Dey et al. [13], for the K_Ga_v of CO_2 absorption into MEA solution in a structured packed column.

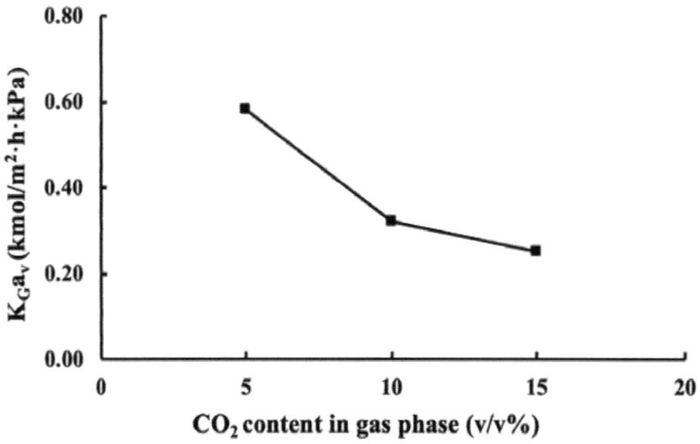

Figure 2. Effect of CO_2 content in gas phase on K_Ga_v. Solvent concentration = 3 kmol/m^3; solvent flow rate = 10.6 m^3/m^{2}h; CO_2 content in gas phase = 5-15 v/v%; CO_2 loading = 0.0 mole CO_2/mole amine.

3.3. Effect of CO_2 initial loading

For the effect of CO_2 initial loading of MEA, it can be found in Figure 3. CO_2 initial loading is defined as mole of CO_2 absorbed per mole of amine due to the absorptions [10]. The results show that the CO_2 initial loading at 0.0 mole/mole give the highest K_Ga_v and 0.2 mole/mole is the lowest K_Ga_v. The CO_2 loading from the liquid in the bottom of the column over range of 0.20-0.35 mole/mole. MEA can absorb the CO_2 around 0.5 mole/mole with absorption process [11]. There are more fresh-amine molecules in amine solution with 0.0 mole/mole for reacting with dissolved CO_2 whereas CO_2 initial loading at 0.2 mole/mole could provide less amine molecules for reacting with CO_2. Therefore, an increase in CO_2 loading within amine solution leads to after decrease amount of fresh amine molecules and K_Ga_v because of a driving force from the gas to the liquid phase will decrease.

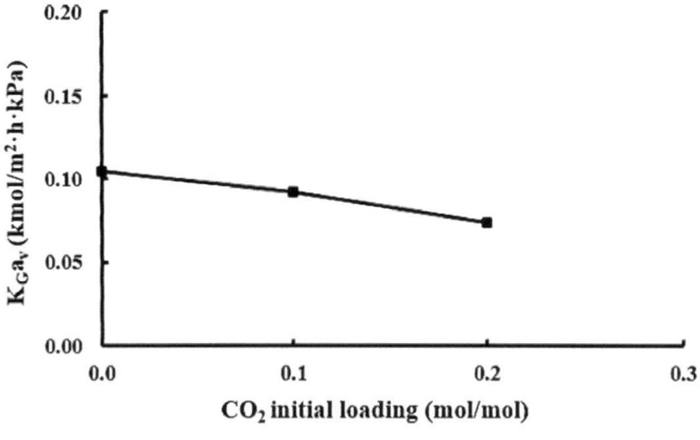

Figure 3. Effect of CO_2 initial loading on K_Ga_v. Solvent concentration = 3 kmol/m^3; solvent flow rate = 10.6 m^3/m^{2}h; CO_2 content in gas phase = 15 v/v%; CO_2 loading = 0.0-0.2 mole CO_2/mole amine.

4. Conclusion

Overall mass transfer coefficient on the CO_2 absorption in MEA solution is experimentally determined by using a random-packed absorption column. The operating conditions are conducted by solution flow rate, CO_2 content in gas phase and CO_2 initial loading. The results show that the mass transfer increase as solution flow rate increases whereas decreases as CO_2 initial loading increase. It can be concluded that 15.9 $m^3/m^2{\cdot}h$ is the best CO_2 absorption condition because there are free amine molecules to react with CO_2 molecules more than other solution flow rate. However, an increase solution flow rate effect to high energy and high cost in the system.

Acknowledgments

The authors would like to gratefully acknowledge the Malaysia-Thailand Joint Authority (MTJA) for financial support of this project.

References

[1] EPA 2017 *United States Environmental Protection Agency*
[2] Kim S, Shi H and Lee J 2016 *International Journal of Greenhouse Gas Control* 45 186
[3] Wang Y, Zhao L, Otto A, Robinius M and Stolten, D 2017 *Energy Procedia* 114 650
[4] Welty J R, Wicks C E, Wilson R E and Rooror G L *Fundamentals of momentum, heat, and mass transfer*. 2009: John Wiley & Sons.
[5] Maneeintr K, Idem O R, Tontiwachwuthikul P and Wee G H A 2010 *Industrial & Engineering Chemistry Research* 49 2857
[6] Halim H N A, Shariff A M, Tan L S and Bustam M A 2015 Industrial & Engineering Chemistry Research 54(5) 1675
[7] Naami A, Sema T, Edali M, Liang Z, Idem R and Tontiwachwuthikul P 2012 *Industrial & Engineering Chemistry Research* 51(18) 6470
[8] Jintai metal raschig ring 2005 China Jintai. Retrieved June 30, 2019, from www.ceramic-honeycombs.com/tower_packing/Raschig_Ring_Tower_Random_Column_Packing
[9] Horwitz W Association of Official Analytical Chemists (AOAC) Methods, 12th ed.; George Banta: Menasha, WI, 1975.
[10] Wen L, Liu H, Rongwong W, Liang Z, Fu K, Idem R and Tontiwachwuthikul P 2015 *Chemical Engineering & Technology* 38 1435
[11] deMontigny D, Tontiwachwuthikul P, Chakma A 2001 *The Canadian Journal of Chemical Engineering* 79 137
[12] Morteza A and Masoud M 2017 *RSC Advances* 7(29) 17857
[13] Dey A and Aroonwilas A 2009 *Energy Procedia* 1 211

IOP Conf. Series: Materials Science and Engineering **859** (2020) 012012 doi:10.1088/1757-899X/859/1/012012

Glycerin Separation from Biodiesel Transesterification Process by Pulsed Electric Field with Specific Pulse Forming Network

Thanadol Hinthao[1,a], **Tanakorn Wongwuttanasatian**[2,b], **Warayut Kampeerawat**[1,c], **Amnart Suksri**[2,d]

[1] Department of Electrical Engineering, Faculty of Engineering, Khon Kaen University, Khon Kaen, Thailand.
[2] Center of Alternative Energy Research and Development, Faculty of Engineering, Khon Kaen University, Khon Kaen, Thailand

[a]Thanadol_h@kkumail.com, [b]tanwon@kku.ac.th, [c]kampeeyut@gmail.com, [d]samnar@kku.ac.th

Abstract. Biodiesel production process can be rapidly done if the glycerin separation can be removed faster. With the palm oil crisis, biodiesel is needed for faster production to add more value and to solve the oversupply problem. Pulse forming network circuit can generate pulsed electric field (PEF) to speedily separate glycerin from biodiesel production. While the substance is reacting, the electrical impedance value of glycerin is changed, the pulse forming network will keep waveform to be a square wave. Transesterification process using palm oil mixed with methanol with a molar ratio of 1:6 by using 1 wt.% of KOH as a catalyst. The reaction chamber electrode was coaxial cylindrical electrodes with diameter 6 cm and 1 cm. The maximum voltage across the reaction chamber is 500 V with 1 ,5 and 10 kHz frequency. Glycerin separation was best achieved when using 5 kHz frequency. The glycerin was obtained at 155 ml in 20 minutes.

1. Introduction

Presently, Palm plantation have been planted around South East Asia. The raw material of palm oil has oversupply and the price has fallen recently. To make full use of the palm oil product and to overcome the palm oil crisis situation, biodiesel is most preferred product that can help increase the value of palm and palm oil industry. It is also by means of a replacement of the usage of fossil fuel that cause the environment problem.

Most common method used in producing biodiesel is a transesterification process. Palm oil reacts with alcohol and a base-catalyzed at 40-60° C[1]. The product obtained from process is biodiesel, glycerin and water as a byproduct. The yield of biodiesel depends on the molar ratio of the substrate. While the process is reacting, the substances will separate from each other by gravity acceleration. After the process, the biodiesel is then washed by water to remove the alcohol and glycerin[2].

To separate the biodiesel and glycerin, the gravity method takes more than 2-4 hours[3] to separate the substances. The chemical process can reduce the separation time but the chemical such as the deep eutectic solvent (DES)[4] makes the biodiesel contaminated. The centrifuge machine is most used method to separate glycerin in oil industry but it is difficult for a small industry due to the expensive

Content from this work may be used under the terms of the Creative Commons Attribution 3.0 licence. Any further distribution of this work must maintain attribution to the author(s) and the title of the work, journal citation and DOI.
Published under licence by IOP Publishing Ltd

price of the machine and high energy consumption. The method that using heat generated such as boiling and ultrasonic[5-6] at temperature of more than 60° C reduce the yield of biodiesel because of alcohol evaporation[7].

Electric field method with low current can shorten the separation time that depends on electric field stress, type of electrode[8] and the frequency of electric field. The goal achievement is to sediment the glycerin in fluid form and to reduce the viscosity faster[9-10].

2. Electric field and glycerin molecule

Electric field affects the polar molecule according to the field direction and waveform length. For alternating current (AC) electric field, polar molecule as glycerin will arrange in the direction of electric field that polarized between positive and negative voltage. The rotation of molecule will create a chance of collisions and pass on energy between molecule[11].

The direct current (DC) electric field will trapped glycerin molecule in the direction of the electric field. Glycerin molecule will then be aligned to the crystalline chain and created dipole bond between molecule[12], the dipole bond will result to larger particle size. However, molecule that moves in perpendicular direction to electric field has decreased the velocity while travelling into DC electric field[13].

As for a pulse electric field (PEF), glycerin molecule will align according to the same direction as for DC electric field. But when an electric field is not energized, the glycerin molecule will sediment to the bottom faster because of the larger size of glycerin molecule. The time that glycerin coagulates and sediment depends on the on/off states of electric field duration. All of electric field applying methods and molecule alignment are shown in Figure 1.

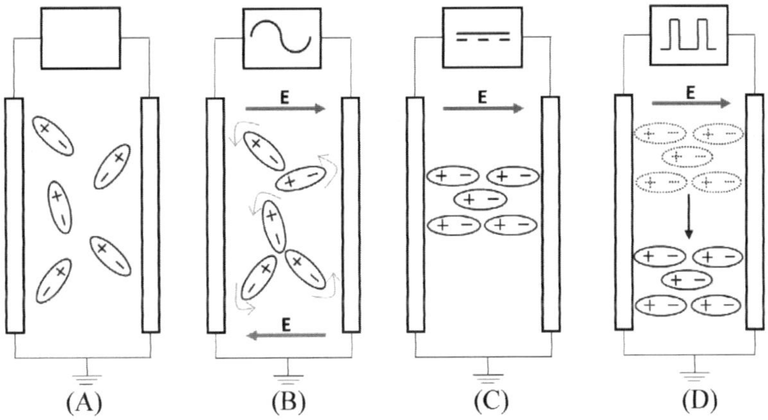

Figure 1. Glycerin molecule in (A) no E-field, (B) AC E-field, (C) DC E-field, (D) PEF.

3. Material and method

In transesterification process, palm oil was prepared by heating at 55° C as substrate. Palm oil and alcohol with a molar ratio of 1:6 was mixed with 1 wt. % potassium hydroxide (KOH) as catalyst, it was reacted until became Homogeneous liquid. Mixed the all substance and stirred for 1 minute, it will change to biodiesel and glycerin. After the transesterification reaction occurred, apply the PEF to the reaction chamber. The chamber made of glass that attached with valve. The valve situated at the bottom of chamber used to export glycerin that slit to the bottom of the chamber. The diameter of designed ground electrode was 6 cm. High voltage electrode diameter was 1 cm and both electrodes had a length of 50 cm. The electrode is sectionalized by 5 cm apart for 5 sections to avoid the trapped molecule and to observe the glycerin particle.

To produce the square pulse waveform, PEF is generated from pulse forming network circuit as shown in Figure2. Pulse forming network will charge the capacitor while switch has closed. After switch on via switching mosfet, the charges from capacitor will be discharged through the inductor

IOP Conf. Series: Materials Science and Engineering **859** (2020) 012012 doi:10.1088/1757-899X/859/1/012012

sections. The length and shape of waveform depends on the impedance of pulse forming circuit[14-15] and load. The capacitor 0.47 μF and inductor for 5 mH for each section has an impedance of 110 Ω. The maximum voltage applied was at 500 V across the chamber. Using higher voltage can generate the breakdown through the glycerin particle and may lead to ignition because glycerin molecules become more conductive.

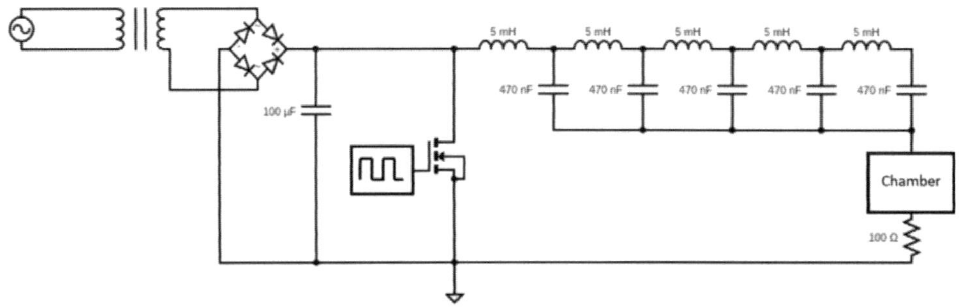

Figure 2. Pulse forming network circuit diagram.

The high voltage switch used in this experiment is high voltage mosfet (IXTF1N450) and it was controlled by TL494 as a trigger signal with 1 ,5 and 10 kHz frequency. The current that flow through the substance was unsteady at first because of the glycerin mixture. To measure the current and the impedance of substance, the resistance of 100 Ω (R_0) was connected series to the chamber. The waveform is then measured by high voltage probe (FLUKE 80K-40) and oscilloscope (Tektronix TDFS1001B). The device configuration is shown in Figure 3.

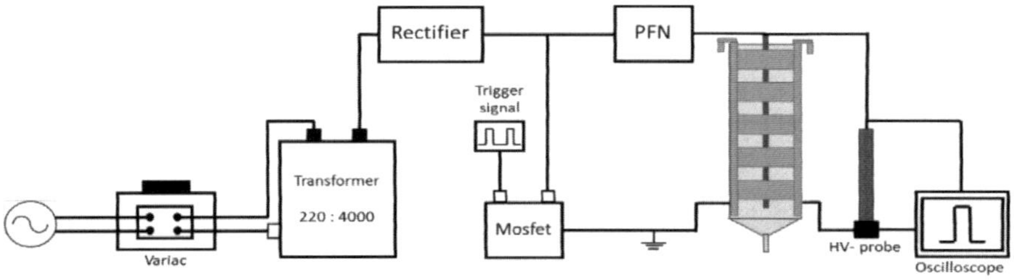

Figure 3. Pulse electric field generator equipment with PEF Chamber.

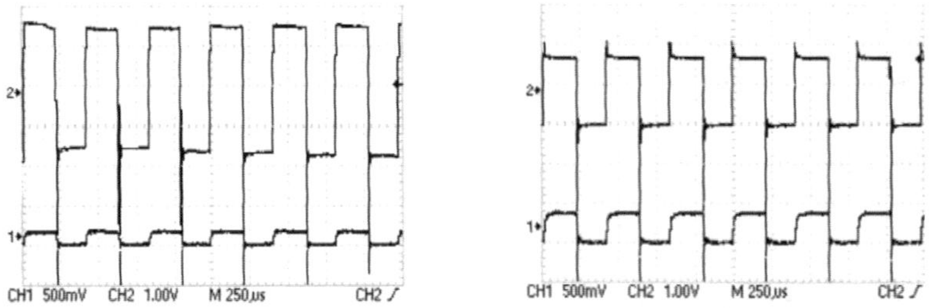

(A) (B)

Figure 4. Voltage across chamber multiplied by 1000 (1) and across R0 multiplied by 10 (2) at (A) start applying and (B) after 5 minutes.

The waveform that measured from R0 has shown the characteristic impedance of the substance and the current that flow through it. The magnitude of voltage has changed while PEF was applying. At first, the waveform showed the voltage at start applying and after 5 minutes in Figure4. The voltage across the chamber is increased due to an increasing of the substance impedance.

4. Results

Pulse electric field can cause the combination process of glycerin. After applying PEF, larger particle size of glycerin molecule is formed, and sedimented to the bottom of the chamber rapidly as shown in Figure 5. The 150 ml glycerin is obtained from transesterification process using 1000 mL of palm oil.

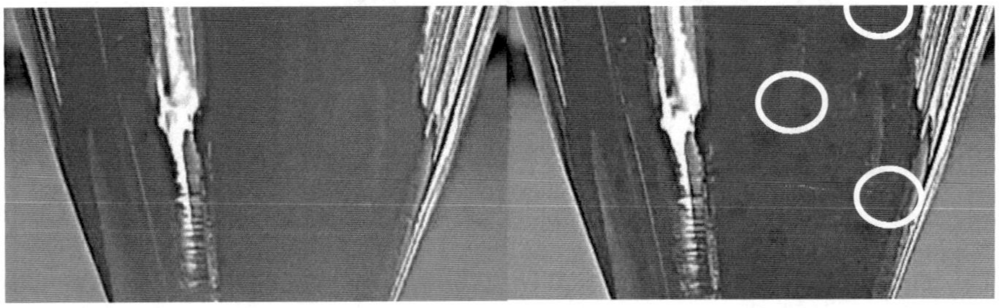

(A) **(B)**

Figure 5. Glycerin (A) before applying PEF and (B) during applying PEF.

The observe time was selected for 5, 10, 15 and 20 minutes for 1 kHz, 5 kHz, 10 kHz and observe heating method with temperature of 60° C. the amount of glycerin obtained is shown in Table1.

Table 1. Amount of glycerin compared from each process.

Observe time	Volume of glycerin (mL)			
(min)	60 °C	1 kHz	5 kHz	10 kHz
5	15	120	130	100
10	23	135	145	125
15	30	145	150	133
20	37	148	155	138

Glycerin is rapidly sedimented from 5 to 10 minutes and slow down afterward. This is due to the amount of glycerin that mixed with biodiesel has been decreased, the remaining glycerin particle are spread-out and has less opportunity for a recombination process.

Table 2. Impedance changing from each process.

Observe time	Impedance of substance [Ω]		
(min)	1 kHz	5 kHz	10 kHz
5	666.67	571.43	416.67
10	1666.67	4375	1666.67
15	2500	4375	2222.22
20	3333.33	4375	2500

The electrical impedance of substance has increased during process because the glycerin that already sedimented to the bottom of the chamber. The remaining substance contained most of the biodiesel product that have poor conductivity. The impedance of substance that changes during process of applying PEF is shown in Table2.

5. Conclusion

Glycerin separation from biodiesel transesterification process by Pulsed Electric Field with specific pulse forming network have been investigated. The pulse forming network will generate pulsed electric field and keep waveshape to the square wave even if the impedance of substance has changed. The PEF can increase the rate of glycerin separation at an initial of the process until 10 minutes. Using pulse forming network with the frequency of 5 kHz process, most glycerin volume were obtained. PEF with 10 kHz process has sedimented slowly because of high frequency waveform switching behaves as DC electric field. It can be concluded that, while the electrical impedance of the substance was increasing slowly, the voltage across the substance has also been increased. Hence, Glycerin separation using pulse electric field and pulse forming network can be a promising technique for a biodiesel industry in the near future.

Acknowledgments
This research project is fully supported by Center for Alternative Energy Research and Development, Khon Kaen University, Thailand.

References

[1] A. A. Mayvan et al., "Current Biodiesel Production Technologies : a Comparative Review," Most, vol. 63, pp. 4–7, 2011.
[2] S. B. G. Ć and D. U. Skala, "DESIGN AND OPTIMISATION OF PURIFI- CATION PROCEDURE FOR BIODIESEL WASHING *," vol. 15, no. 3, pp. 159–168, 2009.
[3] G. Anastopoulos, Y. Zannikou, S. Stournas, and S. Kalligeros, "Transesterification of vegetable oils with ethanol and characterization of the key fuel properties of ethyl esters," Energies, vol. 2, no. 2, pp. 362–376, 2009.
[4] M. Hayyan, F. S. Mjalli, M. A. Hashim, and I. M. AlNashef, "A novel technique for separating glycerine from palm oil-based biodiesel using ionic liquids," Fuel Process. Technol., vol. 91, no. 1, pp. 116–120, 2010.
[5] I. Istadi, A. D. Yudhistira, D. D. Anggoro, and L. Buchori, "Electro-catalysis system for biodiesel synthesis from palm oil over Dielectric-Barrier Discharge plasma reactor," Bull. Chem. React. Eng. Catal., vol. 9, no. 2, pp. 111–120, 2014.
[6] A. B. Koc and E. H. McKenzie, "Effects of ultrasonication on glycerin separation during transesterification of soybean oil," Fuel Process. Technol., vol. 91, no. 7, pp. 743–748, 2010.
[7] N. Sirirat, Biodiesel Production From Transesterification Process Using Electric Field. Chiang Mai: Graduate School , Chiang Mai University, 2011.
[8] A. Abbaszadeh, B. Ghobadian, and G. Najafi, "Electrostatic coagulation for separation of crude glycerin from biodiesel," Adv. Environ. Biol., vol. 8, no. 1, pp. 321–324, 2014.
[9] Y. Hemar, "The effect of pulsed electric field processing on particle size and viscosity of milk and milk concentrates," no. May 2014, 2011.
[10] R. Tao et al., "Reducing the Viscosity of Crude Oil by Pulsed Electric or Magnetic Field," no. 7, pp. 2046–2051, 2006.
[11] T. Ratanabuntha, K. Tonmitr, and A. Suksri, "Acceleration in biodiesel production from palm oil process by high voltage electric field," pp. 225–230, 2018.
[12] E. Abedi, M. A. Sahari, M. Barzegar, and M. H. Azizi, "Designing of high voltage electric field for soybean and sunflower oil bleaching," Innov. Food Sci. Emerg. Technol., vol. 36, pp. 173–180, 2016.
[13] H. L. Bethlem, F. M. H. Crompvoets, R. T. Jongma, S. Y. T. van de Meerakker, and G. Meijer, Deceleration and trapping of ammonia using time-varying electric fields, vol. 65, no. 5. 2002.
[14] J. Saisut, "Design and Low Power Test of Pulse Forming Network for Klystron Modulator at Chiang Mai University," Energy Procedia, vol. 89, pp. 104–109, 2016.
[15] M. Wofford, M. C. Baker, and M. Day, "Design, Simulation, And Testing Of A Pulse Forming Network- Transformer Power Supply For The Texas Tech Railgun," pp. 138–140, 2005.

IOP Conf. Series: Materials Science and Engineering **859** (2020) 012013 doi:10.1088/1757-899X/859/1/012013

The Potential of Oil Palm Ash and Eggshell Powder as Hybrid Biofillers in Natural Rubber Biocomposites

Philip Tan Shien Ming[1], Chai Ai Bao[1*], Shamsul Kamaruddin[2], Ch'ng Shiau Ying[3]

[1]Department of Mechanical, Materials and Manufacturing Engineering, University of Nottingham Malaysia, Selangor, Malaysia.
[2]Engineering Design and Product Development, Technology and Engineering Division, Malaysia Rubber Board, Selangor, Malaysia.
[3]Department of Mechanical Engineering, University of Southampton Malaysia, Johor, Malaysia.

aibao.chai@nottingham.edu.my

Abstract. Astronomical amounts of solid biomass wastes are generated globally each day. It is imperative that the potential of conversion and utilisation of these wastes be identified and implemented, thus reducing landfill load and recovering value from this immense source of material. Natural rubber biocomposites were produced by compounding natural rubber matrix with oil palm ash and eggshell powder as hybrid fillers. The mechanical properties of the biocomposites were determined. It is observed that the biocomposites exhibit higher stiffness than the unfilled natural rubber compound. Acid treatment is found to further increase the stiffness of the biocomposites.

1. Introduction

Natural rubber (NR) is a polymer of isoprene (2-methylbuta-1,3-diene) with the chemical formula C_5H_8. NR possesses excellent physical properties such as high mechanical strength, resistance to impact and tear and low heat build-up. However, NR does have its own drawbacks such as low flame resistance and vulnerability to solvents and chemicals, subsequently limiting its applications. As such, this drawback is typically resolved by mixing the NR with petroleum-based fillers such as carbon black to enhance the mechanical properties of the rubber.

Food industries produce a substantial amount of solid waste each day. Consequently, it is becoming more and more imperative to curb this rapidly expanding issue. With this in mind, optimised systems for food waste management are turning into a necessity given the increasingly stringent constraints pertaining to the environmental issues [1]. Among the numerous solid waste materials available, eggshell powder (ESP) and oil palm ash (OPA) were chosen to be the focus of this paper as potential NR bio-fillers. In fact, these materials have yielded encouraging results when incorporated separately into the NR matrix in previous research works [2, 3, 4, 5]. OPA incorporation yielded the improved cure rate index and increased the elongation at break by 7.4% [4]. Ooi et.al [3] found that the degree of swelling decreased with OPA loading and this was further reduced with the hydrochloric acid (HCl) treatment of the OPA. Moreover, an improvement in crosslink density and improved rubber-filler interaction were observed for HCl-treated OPA in comparison to the untreated OPA composites. The

Content from this work may be used under the terms of the Creative Commons Attribution 3.0 licence. Any further distribution of this work must maintain attribution to the author(s) and the title of the work, journal citation and DOI.
Published under licence by IOP Publishing Ltd

IOP Conf. Series: Materials Science and Engineering **859** (2020) 012013 doi:10.1088/1757-899X/859/1/012013

properties such as load bearing capacity, aging resistance and dynamic mechanical properties improved as well. The thermal stability of OPA-filled NR composites was maintained even for the post-treated samples. For the ESP composites, it was found that the tensile strength and elongation decrease with increase ESP filler loading. On the contrary, the modulus of elasticity and hardness increase as the ESP filler loading increases [6, 7].

As a food item, eggs are regarded as an essential ingredient in the diets of people everywhere in the world. This can be attributed to the high nutritional content and ease of preparation of the eggs as food. ES waste is often disposed of as food waste and is listed worldwide as one of the worst environmental problems [2]. Besides that, Yi et al. [8] suggests that the abundance and chemical composition of ES makes it a potential source of filler for lightweight, cost-saving and low-load bearing composites.

As one of the leading palm oil producers in the world, Malaysia generates a substantial amount of oil palm wastes, leading to an alarming excess of biomass waste. Consequently, the oil palm biomass waste significantly raises the expenditures required for its transport and disposal [9]. The oil palm biomass primarily comprises of empty fruit bunches, oil palm fibres & shells, trunks and fronds [10]. These are combusted as an additional source of fuel to operate the steam boilers within the vicinity of the oil palm mills. In this manner, the burned waste post-combustion coined as OPA is obtained. Afterwards, the OPA is typically disposed as landfill [4]. In a particulate-filled NR system, the interfacial adhesion between the OPA and the NR matrix holds a vital role in the enhancement of the desirable mechanical properties [3].

The incorporation of OPA-ESP hybrid fillers in NR matrix has not been explored in previous research works and studies. Nevertheless, the objectives of this project are to develop NR biocomposites and to identify the mechanical properties of the biocomposites. Furthermore, the properties of the biocomposites were compared with unfilled NR and conventional NR filled with carbon black.

2. Materials and experimental methods

2.1. Bio-fillers preparation and characteristic

The raw eggshell was washed thoroughly. Next, the cleaned eggshell was crushed and the pieces were stirred in beaker containing distilled water using mechanical stirrer at an agitation speed of 1100 rpm for an hour to eliminate the unwanted eggshell membrane. The eggshell was then rinsed to remove the residue membrane and dried for 18 hours in a furnace at 100° C. Next, the dried eggshell was milled using an ultra-centrifugal mill with a 500 μm mesh sieve at 6000 rpm. The milled eggshell obtained was then heat-treated at 600° C for 2 hours in a furnace to produce the ESP [5].

The raw OPA was milled into fine powder using the ultra-centrifugal mill with 500 μm sieve at 6000 rpm. The milled OPA was then dried in a vacuum oven at 80° C and 200 millibar for 24 hours to expel moisture [4]. Next, acid-treated OPA was prepared by mixing the powdered OPA in 10% aqueous HCl solution and stirring it with a glass rod for two hours at room temperature. Next, the soaked OPA particles were filtered and washed thoroughly with distilled water. The acid-treated OPA was then dried for 24 hours at 80° C, followed by activation at 105° C in an oven for an hour [3]. Lastly, the HCl-treated OPA was cooled to room temperature and sieved prior to being kept in desiccator.

The crystalline phase and chemical composition of the fillers were identified using the PANalytical X-Ray Diffraction (XRD) instrument with the 2θ range being set between 10° and 90°, with step size and step time of 0.02° (2θ) and 0.5 seconds respectively.

2.2. Biocomposite samples preparation

Seven different rubber compounds were mixed and prepared using the conventional laboratory-sized two roll mill. 500 gram slabs of SMR L rubber were used for the formulation of each of the seven NR compounds. Each ingredient was weighed using digital mass balance according to the rubber formulations indicated in Table 1. The total mixing time was set to be less than 40 minutes in order to prevent the occurrence of pre-mature vulcanisation as a result of the generated excess heat during the compounding process. The nip gap, mill roll speed ratio, time and temperature of mixing, number of passes and sequence of ingredient addition were retained for each mixing to ensure consistency of the process flow for each rubber compound produced.

Table 1. Formulation table for the rubber compounds.

Ingredients added (phr)	NR Gum	A	B	C	CB	D	E
NR (SMR L)	100	100	100	100	100	100	100
Zinc Oxide	5	5	5	5	5	5	5
Stearic Acid	1	1	1	1	1	1	1
CB N550	-	-	-	-	30	-	-
Eggshell powder	-	10	20	30	-	10	30
OPA	-	1	1	1	-	-	-
Acid-treated OPA	-	-	-	-	-	1	1
Permanax TQ (TMQ)	1	1	1	1	1	1	1
Sulphur	2.5	2.5	2.5	2.5	2.5	2.5	2.5
CBS	0.5	0.5	0.5	0.5	0.5	0.5	0.5

2.3. Measurement of mechanical properties

Strip cutouts (width of 6 mm) of the rubber samples for each rubber composite sample were tested using the Instron 5582 Universal Testing Machine. The cross-head speed was set at 500 mm/minute. The gauge length was 25 mm.

3. Results and discussion

3.1. Characterization of fillers

The XRD pattern of ESP was presented in Figure 1. It was found that these XRD patterns exhibit similar crystalline peaks as that of commercial calcium carbonate (CaCO3). The standout peak is found to be around 30° (2θ) which is a common characteristic of crystalline calcite by the reference code 98-007-8903. Hence, this was identified as a structure dominated by crystalline calcite content. The result from XRD was highly consistent with previous works [5, 11]. The XRD pattern of the OPA shown in Figure 2 exhibits a sharp peak with the greatest height at 26.5295° (2θ), which is identified as the compound of graphite 2H matched by the reference code 98-005-2916.

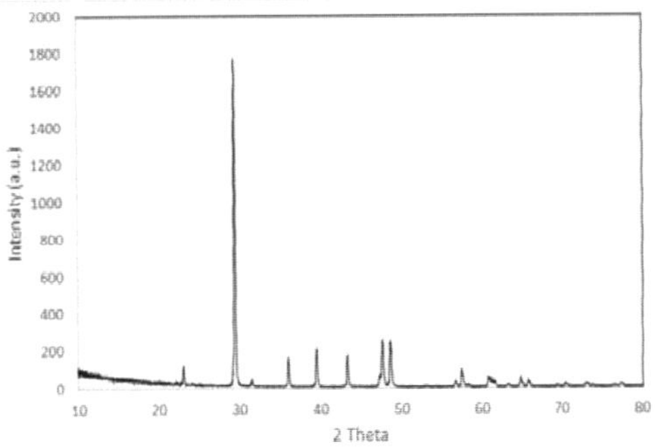

Figure 1. XRD pattern of the ESP sample.

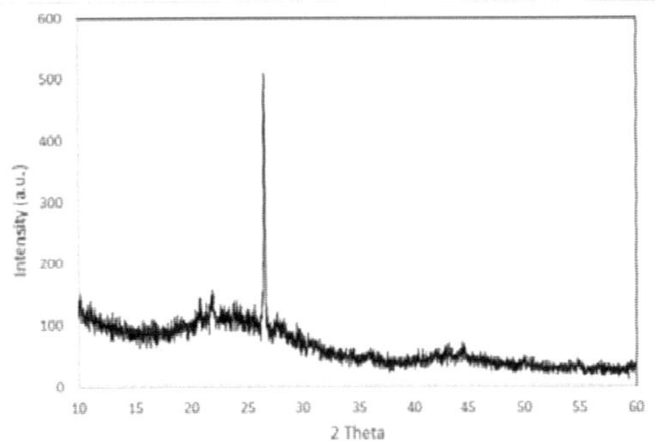

Figure 2. XRD pattern of the OPA sample.

Similarly, the XRD pattern of the acid-treated OPA exhibits the highest peak at 26.4792° (2θ), with several other sharp peaks with lower heights at around 20°, 50°, 67° and 84°. The main compound identified from both sample is quartz as indicated by the highest peak [12]. Both of these samples are near 100% crystallinity based on the absence of 'broad' peaks, with the XRD patterns exhibiting only 'sharp' peaks.

3.2. Mechanical properties

The mechanical properties such as the tensile strength and elongation at break measured from tensile test are illustrated in Figures 3 and 4 respectively. Overall, the observation suggested that the tensile strength for the biocomposites decreases with increasing ESP. This observation can be attributed to the poor interfacial adhesion between the elastomeric matrix and the bio-fillers. The interfacial adhesion serves as an important medium that enable a portion of stress to be transferred from the filler particles during deformation [13]. In addition, filler particles may form aggregates as the amount increases, leading to poorer interfacial adhesion and thus lower tensile strength.

Similarly, the elongation at break decreases with increasing filler content. The addition of filler particles increases the stiffness and hardness of the biocomposites. Subsequently, this results in a reduction of the biocomposites' resilience and toughness and therefore lower elongation at break [14]. As indicated by the reduction, the poor interfacial adhesion reduces the capability of the filler to support possible stress transfer from the bio-fillers to the elastomeric matrix. Therefore, the stress distribution is not uniform and resulting in the biocomposites having poor mechanical properties. This observation found similarities in some other previous works [6, 7, 14] for other combination of matrix-filler systems.

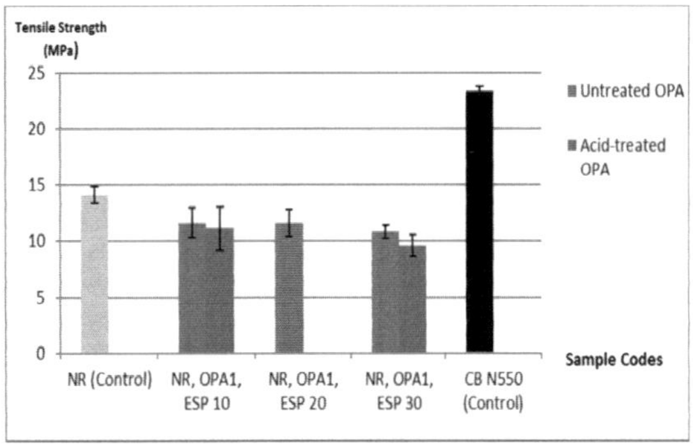

Figure 3. Tensile strengths of the specimens.

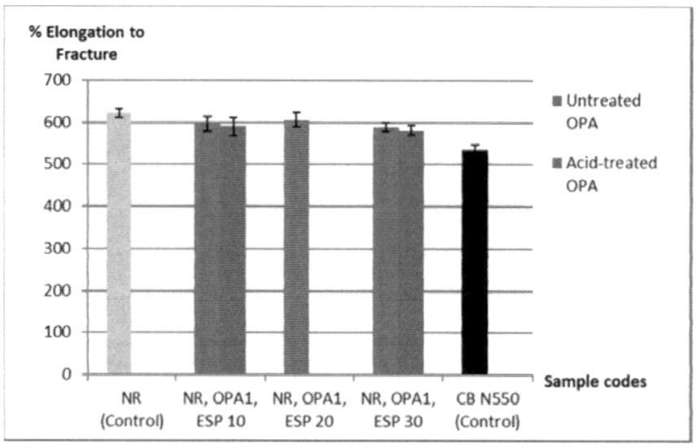

Figure 4. Elongation at break of the specimens.

4. Conclusion

OPA and ESP were incorporated as hybrid bio-fillers to successfully formulate and produce the NR biocomposites. The cure characteristics, mechanical and morphological properties were studied. It was observed that the biocomposites generally show lower tensile strength and lower elongation at break. On contrary, the biocomposites show slightly higher tensile modulus (M100) and higher hardness values. The main factor that affect the mechanical properties is attributed to the interfacial adhesion between the bio-fillers and NR matrix. It was expected that the interfacial adhension could be improved by refining the size of the bio-fillers. These observations indicated that there is a potential to utilize bio-fillers to produce biocomposites in the future.

Acknowledgements

The authors are deeply grateful to the Malaysian Rubber Board for access to the facilities in their Research Centre in Sungai Buloh, Selangor for the mixing and curing of the rubber compounds, in addition to the tensile tests.

References

[1] I. S. Arvanitoyannis, Waste management for the food industries, Academic Press, 2010.

IOP Conf. Series: Materials Science and Engineering **859** (2020) 012013 doi:10.1088/1757-899X/859/1/012013

[2] A. S. Bashir, Y. Manusamy, T. L. Chew, H. Ismail and S. Ramasamy, "Mechanical, Thermal, and Morphological Properties of (Eggshell Powder)-Filled Natural Rubber Latex Foam," Journal of Vinyl and Additive Technology, vol. 23, no. 1, pp. 3-12, 2015.

[3] Z. X. Ooi, H. Ismail and A. A. Bakar, "The Effect of Hydrochloric Acid Treatment on Properties of Oil Palm Ash-filled Natural Rubber Composites," Bioresources, vol. 8, no. 4, pp. 5133-5144, 2013.

[4] Z. X. Ooi, H. Ismail and A. A. Bakar, "Synergistic Effect of Oil Palm Ash Filled Natural Rubber Compound at Low Filler Loading," Polymer Testing, vol. 32, no. 1, pp. 38-44, 2013.

[5] P. Intharapat, A. Kongnoo and K. Kateungngan, "The Potential of Chicken Eggshell Waste as a Bio-filler Filled Epoxidized Natural Rubber (ENR) Composite and its Properties," Journal of Polymers and the Enviroment, vol. 21, no. 1, pp. 245-258, 2013.

[6] S. Shuhadah and A. G. Supri, "LDPE-Isophthalic Acid-Modified Egg Shell Powder Composites (LDPE/ESPI)," Journal of Physical Science, vol. 20, no. 1, pp. 87-98, 2009.

[7] S. Kamaruddin, A. B. Chai, C. L. Lim and J. H. Ho, "Characterisation of static and dynamic responses of natural rubber bio-composites," Journal of Physics: Conference Series, vol. 908, pp. 012-028, 2017.

[8] F. Yi, Z. Guo, P. Hu, Z. Fang, J. Yu and Q. Li, "Mimetics of Eggshell Membrane Protein Fibers by Electrospinning," Macromolecular Rapid Communications, vol. 25, no. 10, pp. 1038-1043, 2004.

[9] Z. X. Ooi, H. Ismail and A. A. Bakar, "Optimisation of Oil Palm Ash as Reinforcement in Natural Rubber Vulcanisation: A Comparison between Silica and Carbon Black Fillers," Polymer Testing, vol. 32, no. 4, pp. 625-630, 2013.

[10] M. F. Awalludin, O. Sulaiman, R. Hashim and W. N. A. W. Nadhari, "An Overview of The Oil Palm Industry in Malaysia and Its Waste Utilization through Thermochemical Conversion, Specifically via Liquefaction," Renewable and Sustainable Energy Reviews, vol. 50, pp. 1469-1484, 2015.

[11] B. Engin, H. Demistas and M. Eken, "Temperature Effects on Egg Shells Investigated by XRD, IR and ESR Techniques," Radiation Physics & Chemistry, vol. 75, pp. 268-277, 2006.

[12] M. Liu, U. Alengaram, M. Santhanam, M. Jumaat and K. Mo, "Microstructural Investigations of Palm Oil Fuel Ash and Fly Ash Based Binders in Lightweight Aggregate Foamed Geopolymer Concrete," Construction & Building Materials, vol. 120, pp. 112-122, 2016.

[13] R. Doufnoune, F. Chebira and N. Haddaoui, "Effect of TItanate Coupling Agent on the Mechanical Properties of Calcium Carbonate Filled Polypropylene," International Journal of Polymeric Materials, vol. 52, pp. 967-984, 2003.

[14] M. Jacob, S. Thomas and K. Varughese, "Mechanical Properties of Sisal/oil Palm Hybrid Fiber Reinforced Natural Rubber Composites," Composites Science & Technology, vol. 64, pp. 955-965, 2004.

[15] D. A. Oliveira, P. Benelli and E. R. Amante, "A Literature Review on Adding Value to Solid Residues: Egg Shells," Journal of Cleaner Production, vol. 46, pp. 42-47, 2013.

IOP Conf. Series: Materials Science and Engineering **859** (2020) 012014 doi:10.1088/1757-899X/859/1/012014

Preparation and Storage Energy Density Base on Dielectric Properties of P(VDF-HFP)/PU/ BNKT Thin Films

K Jehlaeh[1], A Salea[1], P Bomlai[2], and C Putson[1,*]

[1]Department of Physics, Faculty of Science, Prince of Songkla University, Hat Yai, Songkhla, 90110, Thailand
[2]Department of Materials Science and Technology, Faculty of Science, Prince of Songkla University, Hat Yai, Songkhla, 90110, Thailand

*chatchai.p@psu.ac.th

Abstract. Novel electrical devices based on energy storage capacitor and energy conversion, have developed to advanced dielectric properties including electrical performance.In this work, three phases composite of poly(vinylidene fluoride hexafluoropropylene) (P(VDF-HFP)) blended with polyurethane (PU) and filled with $Bi_{0.5}(Na_{0.8}$ -$K_{0.2})_{0.5}TiO_3$ (BNKT) ceramic was studied on the on the dielectric, electrical properties and a storage energy density. All composite thin films are fabricated by the tape casting method in solution. Polarization-Electrical (P-E) loop of all composite thin films were also investigated relaxor-ferroelectric characteristic with the various electric fields. The recoverable energy density and energy efficiency from the loop were then analysed. As a result, the three phases composite of P(VDF-HFP)/PU/BNKT composites produce the highest dielectric constant as well as conductivity; however, they produce the lowest energy efficiency in this work. It seems that the interfacial polarization is easy switchable dipole moment in three phases more than one phase, resulting in high dielectric constant as well as conductivity. However, the larger conductivity seemly generates larger dielectric loss, resulting in lower energy efficiency. Therefore, the comparison based on three and two phases can be used to predict in capability of electric capacitor devices in the future.

1. Introduction

The electric devices such as capacitor, transformer and cables were based on storage energy systems. The development on storage energy density has an important role for improvement of dielectric properties. Most of dielectric materials not only has high dielectric properties, but also has ferroelectric properties that recoverable energy density and high energy conversion efficiency [1]. Normally, there are two types of ferroelectric materials with electrical insulator: ceramic and polymer. For high dielectric properties but small electrical breakdown strength was devoted to ceramic, while the lower dielectric properties and larger electrical breakdown strength was found on polymer. The flexible on dielectric polymer with easy to fabricate and high electrical breakdown strength [2] were considered to main reason for thin film. The combination between ceramic as filler and polymer as matrix was carried out to finding the critical point on storage energy density performance. As reported by Gitanjali [3] and Feng [4], filling ferroelectric polarized ceramic to PVDF polymers is effective on optimizing energy storage by gaining its dielectric properties. In previous paper [5], we found that the three phases of PU which is high electrostrictive property blend with PANi and graphene fillers

Content from this work may be used under the terms of the Creative Commons Attribution 3.0 licence. Any further distribution of this work must maintain attribution to the author(s) and the title of the work, journal citation and DOI.
Published under licence by IOP Publishing Ltd

produced an advanced electrostrictive properties. In this paper, we make another contribution by using that different three phases of PVDF-HFP blend with PU BNKT filler for observing on storage energy density and dielectric properties as well. In this research, P(VDF-HFP) was blend with polyurethane (PU) by DMF solvent before filled with BNKT powder to be thin films. Afterward, final thin films were observed the dielectric properties and storage energy density.

2. Experiments

2.1. *Materials preparation*

Thin films were fabricated by tape casting method with 30 ± 5 μ m thickness. firstly, $Bi_{0.5}(Na_{0.8}$-$K_{0.2})_{0.5}TiO_3$ (BNKT, synthesised by the conventional solid state reactions method) powder was dispersed in N, N-dimethylformamide (DMF, 99% purity, purchased from RCI Labscan Limited, Thailand) by using sonication for 25 min to be the homogeneous solution. Afterward, Polyvinylidene fluoride-hexafluoropropylene (P(VDF-HFP), Solef 11010/1001, purchased from Solvay Solexis, Belgium) powder and thermoplastic polyurethane elastomers (PU, DP 9370A, supplied from Covestro AG, Germany) granules were then dissolved and stirred 60 °C for 5 hours in the solution before removing bubble for 2 hours. Then, the composites solution was cast on the glass plate and dried in the oven at 80 °C for 24 hours. Finally, obtained films were taken off from the smooth glass plate.The pure P(VDF-HFP), P(VDF-HFP)/BNKT and P(VDF-HFP)/PU/BNKT composites were named from the ratio 10:0:0wt% of P(VDF-HFP):PU:BNKT, 10:0:20wt% of P(VDF-HFP)/PU/BNKT, and 8:2:20wt% of PVDF-HFP:PU:BNKT, respectively.

2.2. *Dielectric properties*

Electrical capacitor, dielectric loss on thin films were measured by LCR meter (IM 3533 HIOKI) at frequency range 10^2–10^5 Hz. Then, the dielectric constant and AC conductivity can be calculated from Equation (1) and (2), respectively. Where $\varepsilon_r, C_p, t, A, \varepsilon_o, \sigma_{ac}, f$ and ε_r'' are dielectric constant, electrical capacitor, thickness of film, area of electrode, permittivity of air equalling 8.853 x 10^{-12} F m^{-1}, AC conductivity, frequency, and dielectric loss, respectively.

$$\varepsilon_r = \frac{C_p t}{A\varepsilon_o} \quad (1)$$

$$\sigma_{ac} = 2\pi f \varepsilon_0 \varepsilon_r \varepsilon_r'' \quad (2)$$

2.3. *Ferroelectric P-E loops properties*

Polarization-Electric filed (P-E) loops were measured via a ferroelectric analyser under different electric fields at room temperature. Recoverable energy density (U_e) can be calculated from the curve of P-E loop as Equation (3). For loss energy density (U_l) can be calculated in the closed area in P-E loop [4-6]. Furthermore, energy conversion efficiency (η) was evaluated by using the ratio of U_e and U_l following in Equation (4) [6, 7]. Where U_e, E, P, P_r, P_{max}, and η were recoverable energy density, electric field, polarization, remnant polarization, the maximum polarization, energy conversion efficiency, respectively.

$$U_e = \int_{P_r}^{P_{max}} E dP \quad (3)$$

$$\eta(\%) = \frac{U_e}{U_e + U_l} \times 100 \quad (4)$$

3. Results and discussion

3.1. *Dielectric properties*

All samples were measured for dielectric properties as showed in showed in Figure 1. In Figure 1 (A) showed the relationship between The dielectric constant (ε_r) in various frequencies (10^2 – 10^5 Hz) of pure P(VDF-HFP), P(VDF-HFP)/BNKT and P(VDF-HFP)/PU/BNKT composites. It can be observed that all samples have high dielectric constant at the lowest frequency. The decrease of dielectric constant was observed when the frequency increases because of represents orientational polarizability, matching with prior study [8]. The dielectric constant increases with increasing phases materials from pure PVDF-HFP, two phases P(VDF-HFP)/BNKT, and three phases P(VDF-HFP)/PU/BNKT, respectively. The highest dielectric constant was devoted to 5.70 at 100 Hz for three phases P(VDF-HFP)/PU/BNKT. Similar trend with dielectric loss (ε_r'') and AC electrical conductivity (σ_{ac}), it showed in Figure 1 (B)-(C), respectively. Moreover, Figure 1 (B) showed the dielectric loss in function of frequency. It seems that the dielectric loss (ε_r'') hit a lowest peak at frequency around 10^3-10^4 Hz. The highest loss was seemly presented at the lowest (10^2 Hz) because range is interfacial polarization that easily switchable polarization under electric field. Moreover, the highest frequencies (10^5 Hz) also show the high dielectric loss. The dipole moment may hardly switchable, resulting in high dielectric loss. However, too much dielectric losses were undesirable for electronic devices. In addition, Figure 1 (C) exhibited the AC electrical conductivity (σ_{ac}) in function of frequency. This conductivity was related to the number of charges carries with electrical resistance. It can see that the conductivity was increased with increasing phases materials from pure P(VDF-HFP), P(VDF-HFP)/BNKT and P(VDF-HFP)/PU/BNKT. The three phases materials possible produce the highest charges carries that hop between phases. They gradually increase with frequency as described by the MWS model [9, 10].

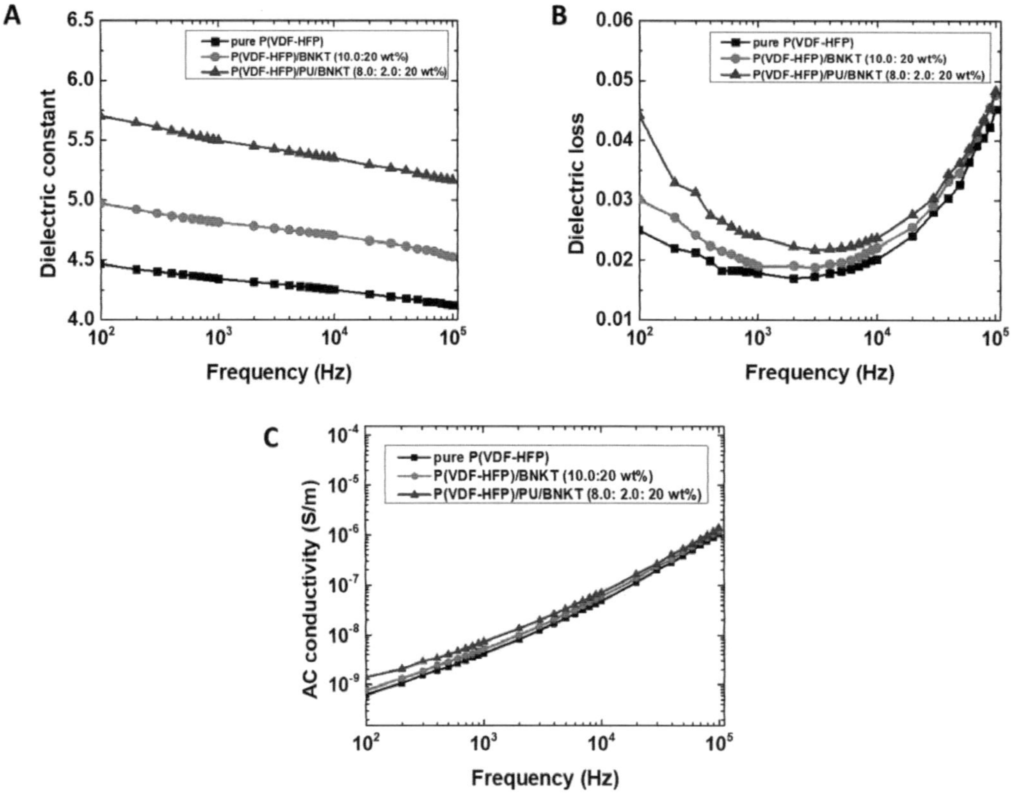

Figure 1. (A) dielectric constant, (B) dielectric loss, and (C) AC conductivity of pure PVDF-HFP, two phases P(VDF-HFP)/BNKT, and three phases P(VDF-HFP)/PU/BNKT thin films at room temperature.

3.2. Ferroelectric P-E loops properties

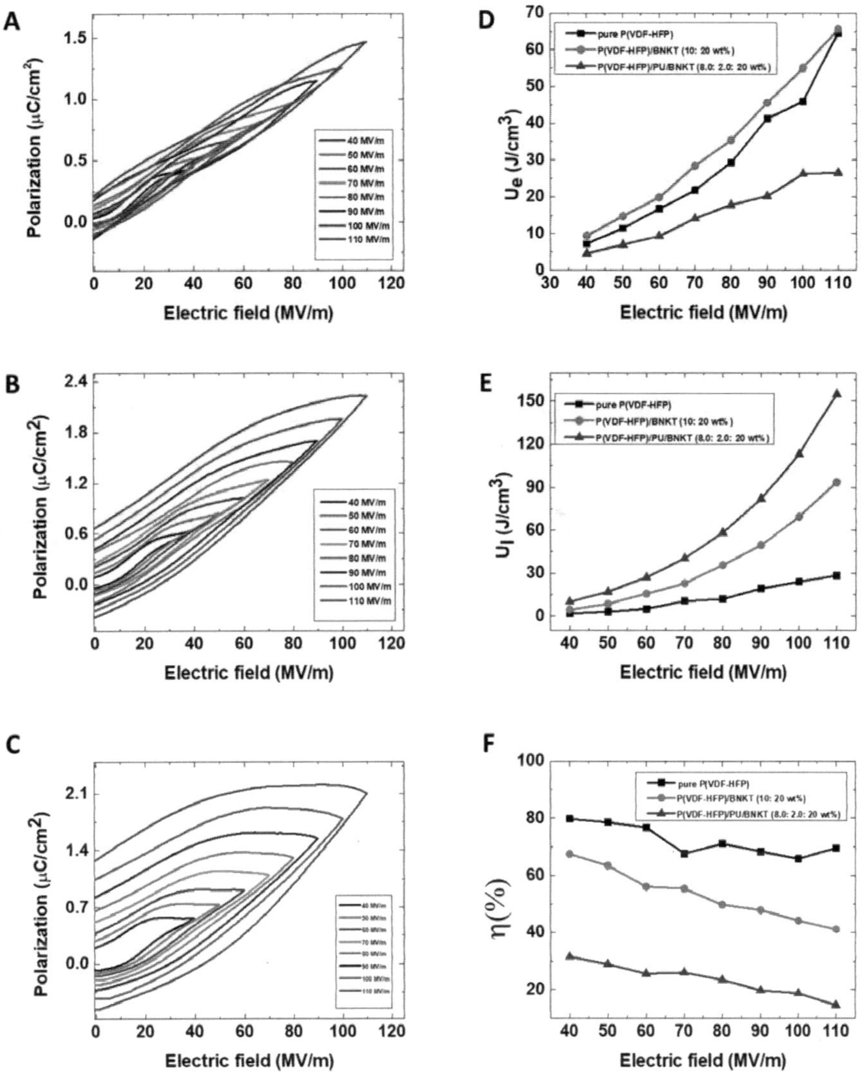

Figure 2. The unipolar P-E loop of (A) pure P(VDF-HFP), (B) P(VDF-HFP)/BNKT composite, (C) P(VDF-HFP)/PU/BNKT composite. (D) recoverable energy density, (E) loss energy density, and (F) energy conversion efficiency of pure P(VDF-HFP), P(VDF-HFP)/BNKT, P(VDF-HFP)/PU/BNKT composites under different electric fields.

From Figure 2 showed the P-E loops and their storage energy density on pure P(VDF-HFP), P(VDF-HFP)/BNKT, and P(VDF-HFP)/PU/BNKT composites. The Figure 2 (A-C) display about the unipolar P-E loop of all samples. They were measured under electric filed from 40-110 MV/m at 10 Hz frequency with room temperature. It can be observed that the characteristic of P-E loops was bigger with applied electric field, resulting in larger P_r and P_{max}. Considering at 110 MV/m, the lowest of P_{max} was belonged to pure P(VDF-HFP); while, the highest of P_{max} was devoted to three phases P(VDF-HFP)/PU/BNKT thin films, equaling 2.10 μ C/cm^2. As a result, the closed area of three phases produce the highest loss energy density as U_l; while, the pure P(VDF-HFP) was the lowest one, equaling 28.35 J/cm^3 as show in Figure 2 (E). This trend of U_l intensification with material phases was seemly related with dielectric loss as show above. The U_l seemly increase in exponential with the external force of electric filed as well as U_e in Figure 2 (D). However, U_e characteristic was slightly difference. It can see that the highest U_e was belonged to P(VDF-HFP)/BNKT; whereas, the lowest U_e was still P(VDF-HFP)/PU/BNKT, equalling 26.54 J/cm^3. In this case, it seems that BNKT was relaxor ferroelectric, resulting in slimmer loop and produce the highest U_e on P(VDF-HFP)/BNKT composite. To more convenient for comparison, the energy conversion efficiency as η was considered in Figure 2 (F). It can see that η tends to decrease with electric field [11], and the lowest η was devoted to three phases P(VDF-HFP)/PU/BNKT. The higher phases possibly produce a large charge carries, resulting in large dielectric loss, U_l, and η, respectively. The movement of interfacial polarization may have easily to move when considered in three phases, resulting in provide the highest dielectric constant [12]. However, the easily movement of the polarization produce the larger energy loss U_l as well. Actually, PU has no ferroelectric properties as P-E loop although it exhibits high dielectric constant, resulting in the smallest η in any electric fields.

4. Conclusion

The thin films of pure P(VDF-HFP), P(VDF-HFP)/BNKT, and P(VDF-HFP)/PU/BNKT composites were fabricated by tape casting method and compared electrical properties such as dielectric properties and ferroelectric P-E loops. The higher phases as three phases P(VDF-HFP)/PU/BNKT composites produce the best dielectric constant, whereas, the three phases provide the lowest ferroelectric P-E loop properties. The pure P(VDF-HFP) is suitable for energy conversion efficiency; while, the two phases P(VDF-HFP)/BNKT is suitable for recoverable energy density because of the relaxor BNKT characteristic. These new three phases relation on dielectric and ferroelectric properties will be able to design the advanced storage energy density in capacitors and electronic devices in future.

Acknowledgments
This work was supported by Thailand Center of Excellence in Physics (ThEP-61-PIP-PSU3), Department of Physics, and department of Materials Science and Technology Faculty of Science, Prince of Songkla University (PSU), Thailand.

References

[1] Zhi-min D, Jin-Kai Y, Sheng-Hong Y and Rui-Jin L 2013 *Adv. Mater.* 25 6334
[2] Yefang F, Qin W, Qihuang D, Cheng P and Zhichao X 2019 *Ceram. Int.* 45 16600-16607
[3] Tabhane G. H., Giripunje S. M. and Kondawar S. B. 2019 *Phys. B (Amsterdam, Neth.)* 571 149-161
[4] Feng Y., Li J. L., Li W. L., Li M. L., Chi Q. G., Zhang T. D. and Fei W. D. 2019 *Composites, Part A* 125 105524
[5] Ardimas, Putson C and Muensit N 2018 *Compos. Sci. Technol.* 158 164
[6] Fei W, Hanya L, Jianfei Y, Wangfeng B, Luwen W, Lili L and Lin Z 2019 *Compos. Sci. Technol.* 182 107759
[7] Pan C, Peng L, Jiwei Z, Bo S, Feng L and Shuanghao W 2017 *Ceram. Int.* 43 13371-13376
[8] Ronald E, Roy D, and Jose P 1998 *Sens. Actuators, A.* 64 77-85
[9] Tsangaris G. M., Psarras G. C., and Kouloumbi N 1998 *J. Mater. Sci.* 33 2027-2037
[10] Salea A, Jehleah K and Putson C 2019 *IOP Conf. Ser. Mater. Sci. Eng.* 625 012011
[11] Luo H, Roscow J, Zhou X, Chen S, Han X, Zhou K, Bowen C. R. 2017 *J. Mater. Chem. A.* 5 7091-7102
[12] Putson C, Jaaoh D and Muensit N 2016 *Mater. Lett.* 172 27-31

IOP Conf. Series: Materials Science and Engineering **859** (2020) 012015 doi:10.1088/1757-899X/859/1/012015

Effect of Heating Process on Flexural Strength and Toughness of CFRTP Molded by Multi-layer Press -Investigation on Effect of Residual Stress-

Akihito ITAMI[1], Kiyotaka OBUNAI[2], Kazuya OKUBO[2], Hiroyuki TAKEI[3]

1Graduate School of Doshisha University, Kyoto, Japan
2Department of Mechanical Engineering, Doshisha University, Kyoto, Japan
3KITAGAWA SEIKI Co., Ltd, Fuchu, Hiroshima

ctwc0520@mail4.doshisha.ac.jp, kiobunai@mail.doshisha.ac.jp, kokubo@mail.doshisha.ac.jp, takei-h@kitagawaseiki.co.jp

Abstract. This study investigated the effect of differences in heating process of each mold due to the molding with multi-layer press on changing in mechanical properties of CFRTP. CFRTP specimens molded with 10 pages multi-layer were prepared for 3 point bending tests and DCB tests, ENF tests in order to investigate the different of mechanical properties. The change in the mechanical properties of CFRTP molded by multi-layer press were investigated considering residual stress occurred by differences in heating process of each mold. Average of flexural strength, mode-I interlaminar fracture toughness and mode-II interlaminar fracture toughness of the specimen molded at 5th page were decreased compared with that of 1st page. Assuming the fiber as a rigid and considering the thermal shrinkage of resin, compressive residual stress should be occurred in the fiber since free shrinkage of the resin was constrained. It was considered that the reduction of flexural strength with the failure close to the compression surface was caused due to this additional compressive residual stress. The tensile residual stress also should be occurred in the resin due to constraint of free shrinkage of the resin. The observed reduction of toughness in the resin was also occurred for the procedure of crack propagation. The reduction rate of mode-I interlaminar fracture toughness was much remarkable than that of mode-II interlaminar fracture toughness. Tensile residual stress in the resin was dominant since the mode-I fracture is a crack opening mode.

1. Introduction

Carbon fiber reinforced thermoplastics (CFRTP) have been attracting attention and widely used in various fields (for example, aerospace engineering, transportation industries, sporting goods and building materials etc.) where weight saving is desired due to their excellence in specific strength and specific modulus as compared with metal materials [1-7]. In order to meet this demand, in recent years, there are strong requirements to establish molding methods that achieves both improved productivity and low cost. In this study, multi-layer press method was applied to molding CFRTP in order to greatly reduce molding time and cost by piling up some molds. Generally, mechanical properties of CFRTP are depended on heating process. In molding CFRTP molded by multi-layer press, since distance of molds from hot plate of the press machine is not constant, heating process of each mold is depended on the set positions in multi-layer. However, the effect of differences in heating process of each mold due to the

Content from this work may be used under the terms of the Creative Commons Attribution 3.0 licence. Any further distribution of this work must maintain attribution to the author(s) and the title of the work, journal citation and DOI.
Published under licence by IOP Publishing Ltd

molding with multi-layer press on changing in mechanical properties of CFRTP has not been clarified. The purpose is to reveal the effect of differences in heating process of each mold due to the molding with multi-layer press on changing in mechanical properties of CFRTP.

In this study, CFRTP specimens molded with 10 pages multi-layer were prepared for 3 points bending tests and DCB tests, ENF tests in order to investigate the different of mechanical properties. The change in the mechanical properties of CFRTP molded by multi-layer press were investigated considering residual stress occurred by differences in heating process of each mold.

2. Experimental program

2.1. Materials

Prepreg sheets of unidirectional continuous carbon fiber fabricated with MXD10 (m-xylylenediamine10: melting point 215 degree Celsius, Young's modulus 3.08GPa, shrinkage strain function 1.1%) as matrix were prepared. The prepreg sheets were supplied by Mitsubishi Engineering-Plastics Corporation, Japan. The volume fraction of carbon fibers was approximately 40% in the prepreg sheets.

2.2. Molding method of multi-layer press

Figure 1 shows the schematic views of multi-layer press. Each page was defined as the unit of molding of multi-layer pressing. The number of pages was 10 for multi-layer press. In this study, each page was called with sequence number in order from the top one with "p" text, so that top and bottom were called (p.1 mold) and (p.10 mold), respectively, as shown in the figure. Prepreg sheets were cut into 750 x 750mm and set into each mold. Figure 2 shows examples of heating process of the 1st, 3rd, 5th and 10th pages during multi-layer pressing. The temperature was elevated at 266 degree Celsius keeping 0.1MPa of pressure. After reaching to the maximum temperature, the temperature was maintained for 60 minutes. After starting cooling, the pressure was increased to 0.5 MPa.

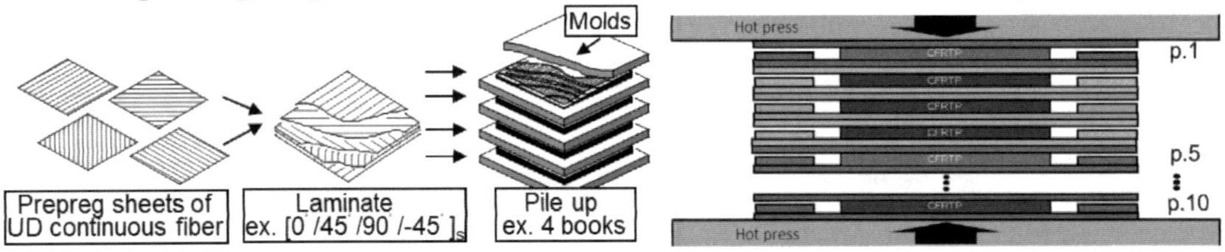

Figure 1. Schematic views of multi-layer press.

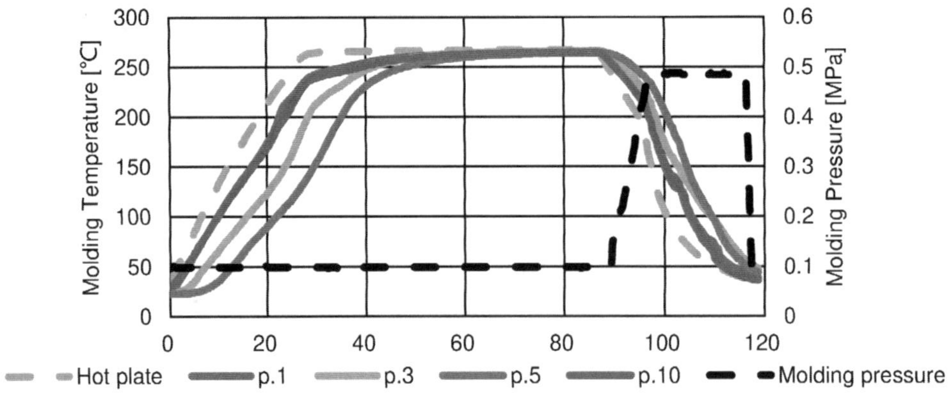

Figure 2. Illustration of molding process diagram.

2.3. Mechanical testing methods

2.3.1. Quasi-static 3 point bending tests

Flexural strength of CFRTP molded by multi-layer press was measured by quasi-static 3 point bending tests according to ASTM D7264 at 5 mm/min (cross head speed) by using a universal material testing machine. For this test, stacking sequence of the specimens was $[0^\circ/45^\circ/90^\circ/-45^\circ]_s$ of quasi-isotropic 8 layers. Dimensions of the specimens were 60 (length) x 15 (width) x 1.2 (thickness) mm. Test data were obtained with at least 5 samples. The states of cross section of specimens after 3 point bending tests were observed by digital microscope.

2.3.2. Double cantilever beam tests

Mode-I interlaminar fracture toughness of CFRTP molded by multi-layer press was measured by double cantilever beam (DCB) tests according to ASTM D5528 at 1 mm/min (cross head speed) by using a universal material testing machine. Figure 3-(a) shows schematic view for DCB tests. For this test, stacking sequence of the specimens was $[(0^\circ/90^\circ)_6/0^\circ]_s$ of orthotropic 26 layers. Aluminum foil with approximately 15μm thickness was previously inserted between 13th and 14th ply of the laminate. Dimensions of the specimens were 250 (length) x 25 (width) x 4.0 (thickness) mm. The length of pre-crack was 50 mm. Test data were obtained with at least 5 samples. Mode-I interlamina fracture toughness was calculated by following equation (1) and (2)[8].

$$G_{IC} = \frac{3}{2(2H)}\left(\frac{P_c}{B}\right)^2 \frac{(BC)^2}{\alpha_1} \tag{1}$$

$$\frac{a}{2H} = a_1(BC)^{\frac{1}{3}} + a_0 \tag{2}$$

where G_{IC}, a, $2H$, B, P_c, C, a_1 and a_0 denote mode-I interlaminar fracture toughness, crack length, specimen thickness, specimen width, initial critical load, loading point compliance of the initial elastic part, the slope of the approximate straight line of plotting $a/2H$ (the dimensionless crack length) on the vertical axis and $(BC)^{1/3}$ (the cubing root of the crack opening displacement compliance per unit width) on the horizontal axis and the intercept of the above approximate straight line, respectively.

2.3.3. End notched flexure tests

Mode-II interlaminar fracture toughness of CFRTP molded by multi-layer press was measured by end notched flexure (ENF) tests according to ASTM D7905 at 0.5 mm/min (cross head speed) by using a universal material testing machine. Figure 3-(b) shows schematic view for ENF tests. For this test, stacking sequence of the specimens was $[(0^\circ/90^\circ)_6/0^\circ]_s$ of orthotropic 26 layers. Aluminum foil was also previously inserted between 13th and 14th ply of the laminate. Test data were obtained with at least 5 samples. Dimensions of the specimesn were 250 (length) x 25 (width) x 4.0 (thickness) mm. The length of pre-crack was 50 mm. Mode-II interlamina fracture toughness was calculated by following equation (3) and (4)[9, 10].

$$G_{IIC} = \frac{9a_1^2 P_c^2 C_1}{2B(2L^3+3a_1^3)} \tag{3}$$

$$a_1 = \left\{\frac{C_1}{C_0}a_0^3 + \frac{2}{3}\left(\frac{C_1}{C_0} - 1\right)L^3\right\}^{\frac{1}{3}} \tag{4}$$

where GIIC, a0, Pc, C0, C1, a1, L and B denote mode-II interlaminar fracture toughness, initial crack length, initial critical load, loading point compliance of the initial elastic part, loading point compliance at initial critical load, estimated crack growth length at initial critical load, supports span and specimen width, respectively.

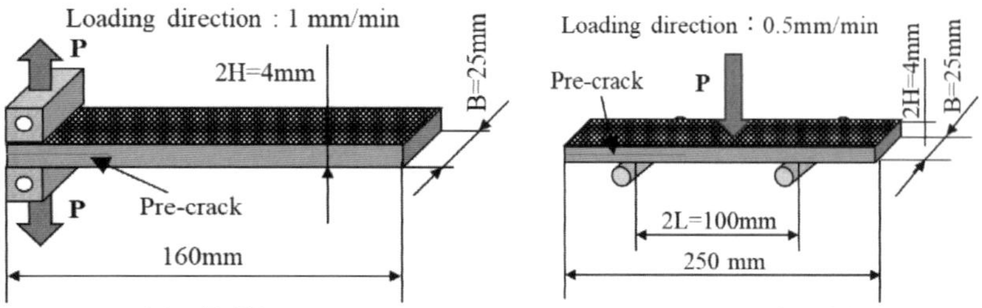

(a) : DCB tests (b) : ENF tests

Figure 3. Schematic views for DCB and ENF tests.

2.4. Estimation of residual stress in carbon fiber

Compressive residual stress should be occurred in the fiber due to the thermal shrinkage of resin was calculated considering linear expansion coefficient of resin. Compressive residual stress was calculated by following Equation (5)

$$\sigma_R = \alpha \Delta t E(1 - V_f) \tag{5}$$

where σ_R, a, Δt, E and V_f denote compressive residual stress should be occurred in the fiber, linear expansion coefficient of resin, temperature difference, Young's modulus of resin and fiber volume fraction, respectively.

3. Results and discussions

3.1. Flexural strength

Figure 4 shows differences of the flexural strengths of the specimens molded at 1st and 5th page. The states of cross sections of specimens after 3 point bending tests are shown in figure 5. Average of flexural strength of the specimen molded at 5th page was decreased approximately 10% compared with that of 1st page. Specimens on each page were failed between layers close to the compression surface and finally broken at the tensile part.

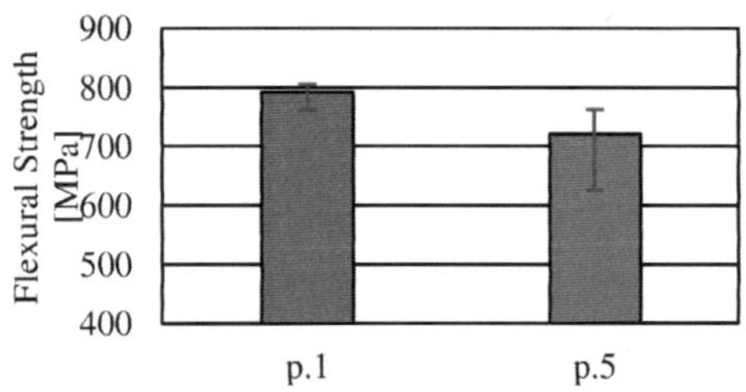

Figure 4. Differences in 3 point bending strengths.

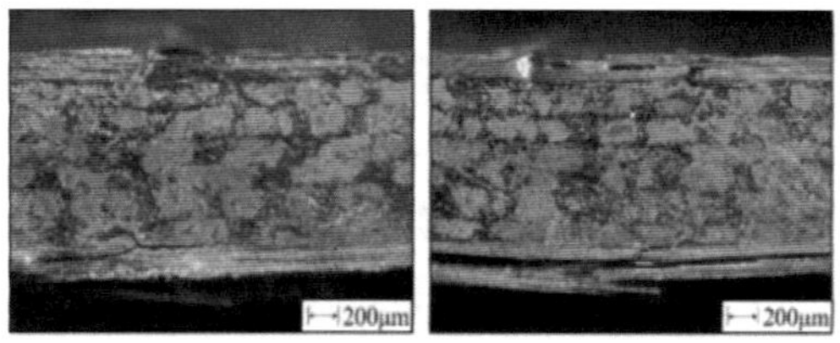

page1 page5

Figure 5. States of cross sections of specimens after 3 point bending tests.

3.2. Mode-I interlaminar fracture toughness

Figure 6 shows differences of the mode-I interlaminar fracture toughness of the specimens molded at 1st and 5th page. Average of mode-I interlaminar fracture toughness of the specimens molded at 5th page was decreased approximately 20% compared with that of 1st page.

3.3. Mode- II interlaminar fracture toughness

Figure 7 shows differences of the mode-II interlaminar fracture toughnesses of the specimens molded at 1st and 5th page. Average of mode-II interlaminar fracture toughness of the specimens molded at 5th page was decreased approximately 13% compared with that of 1st page.

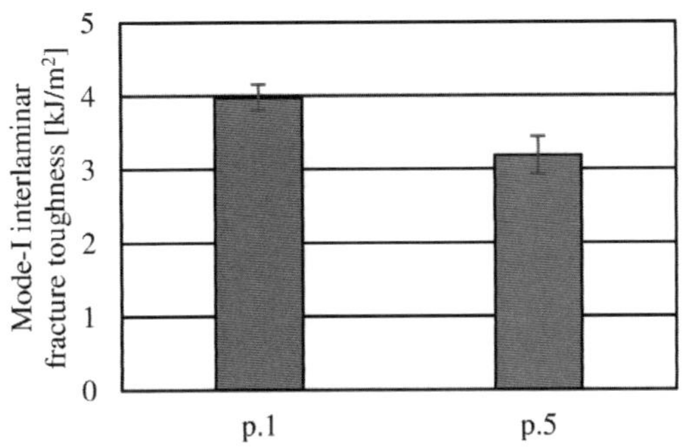

Figure 6. Differences in mode-I interlaminar fracture toughness.

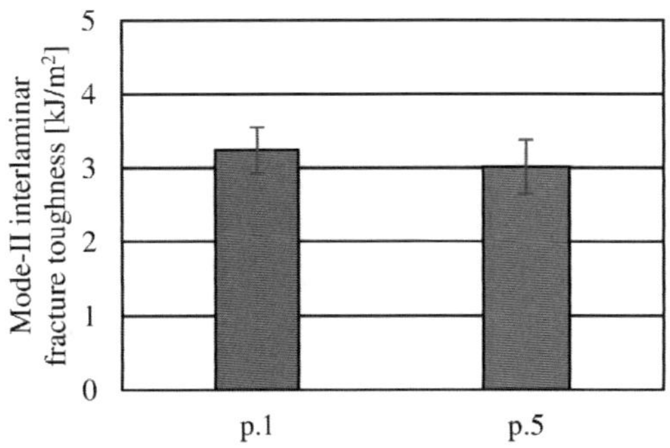

Figure 7. Differences in mode-II interlaminar fracture toughness.

3.4. Effect of residual stress

The temperature in the mold of the 1st page was lower than the melting temperature of the resin, while it was higher than that in the 5th page, during cooling process after molding pressure was respectively reached to 0.5MPa. Figure 8 shows illustrations of the models of residual stress. Assuming the fiber as a rigid and considering the thermal shrinkage of resin, compressive residual stress should be occurred in the fiber since free shrinkage of the resin was constrained. Approximate 20 [MPa] of compressive residual stress of the fiber was estimated according to the differences in heating process, where linear expansion coefficient of resin was assumed to be 55 [10^{-6}/degree Celsius]. It was considered that the reduction of flexural strength with the failure close to the compression surface was caused due to this additional compressive residual stress.

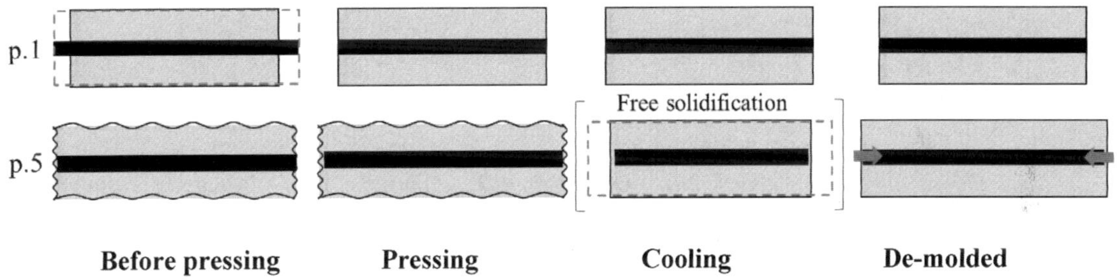

Figure 8. Models of residual stress occurred in fiber.

On contrary, the tensile residual stress also should be occurred in the resin due to constraint of free shrinkage of the resin. The observed reduction of the toughness in the resin was also occurred for the procedure of crack propagation. The reduction rate of the mode-I interlaminar fracture toughness was much remarkable than that of the mode-II interlaminar fracture toughness. Tensile residual stress in the resin was dominant since the mode-I fracture is a crack opening mode.

4. Conclusion

In this study, the effects of heating process on the change of flexural strength and interlaminar fracture toughness of CFRTP molded by multi-layer press in terms of residual stress was discussed. The following conclusions were given.

(1) Average of flexural strength, mode-I interlaminar fracture toughness and mode-II interlaminar fracture toughness of specimens molded at 5th page were decreased compared with those of 1st page.

(2) Approximate 20 [MPa] of compressive residual stress of the fiber was estimated according to the differences in heating process. It was considered that the reduction of flexural strength with the failure close to the compression surface was caused due to this additional compressive residual stress.

IOP Conf. Series: Materials Science and Engineering **859** (2020) 012015 doi:10.1088/1757-899X/859/1/012015

(3) The tensile residual stress also should be occurred in the resin due to constraint of free shrinkage of the resin. The observed reduction of the oughness in the resin was also occurred for the procedure of crack propagation. The reduction rate of mode-I interlaminar fracture toughness was much remarkable than that of mode-II interlaminar fracture toughness. Tensile residual stress in the resin was dominant since the mode-I fracture is a crack opening mode.

References

[1] Osaka Municipal Industrial R & D Laboratory Plastic Reader Editing Committee, Reading Book of Plastics 20th Edition, PLASTICS AGE, (2010)

[2] Hironori Nishida, Valter Carvelli, Toru Fujii, Kazuya Okubo, Quasi-static and fatigue performance of carbon fibre reinforced highly polymerized thermoplastic epoxy, Composites Part B, No.144, (2018), pp.163-170

[3] Mingming Zhu, Min Li, Qing Wu, Yizhuo Gu, Yanxia Li, Zuoguang Zhang, Effect of processing temperarure on the micro- and macro-interfacial properties of carbon fiber/epoxy composites, Composite Interfaces, Vol.21, No.5, (2014), pp.443-453

[4] Hisai Ueda, Wataru Okumura, Hideyuki Uematsu, and Shuichi Tanoue, Processing of Carbon Fiber Fabric Reinforced Polycarbonate -The Influence of Void on the Flexural Property-, Journal of Fiber Science and Technology, Vol.72, No.12, (2016), pp.258-264

[5] Satoshi Kobayashi and Atsushi Tanaka, Resin Impregnation Behavior in Processing of Unidirectional Carbon Fiber Reinforced Thermoplastic Composite, Japan Society for Composite Materials, Vol.38, No.2, 2012, p.51-57

[6] Kazuto Tanaka, Shunsuke Maehata, Tsutao Katayama, Effect of resin supply form on mechanical properties of continuous carbon fiber reinforced thermoplastics, Journal of the Society of Materials Science, Japan, Vol.65, No.8, (2016), pp.592-597

[7] Masachika Yamane, Collection of Carbon Fiber Reinforced Thermoplastic Technology - Materials・Molding・Processing・Recycle-, Science & Technology, (2015)

[8] Yoshiki Furusawa, Hiroshi Saito, Mototsugu Tanaka, Manato Kanesaki and Isao Kimpara, Evaluation of Mode I Static Interlaminar Fracture Toughness in the 0°/45° Layers of CFRP Laminates, Japan Society for Composite Materials, Vol.44, No.4, 2018, p.115-122

[9] Yoshiki Furusawa, Hiroshi Saito, Mototsugu Tanaka, Manato Kanesaki and Isao Kimpara, Evaluation of Mode II Static Interlaminar Fracture Toughness in 0°/45° Oriented Layers of CFRP Laminates, Japan Society for Composite Materials, Vol.44, No.5, 2018, p.195-202

[10] THE JAPAN REINFORCED PLASTICS SOCIETY, FRP Basics -From Basics to Applications of Fiber Reinforced Plasics-, CORONA PUBLISHING CO., LTD., (2016).

IOP Conf. Series: Materials Science and Engineering **859** (2020) 012016 doi:10.1088/1757-899X/859/1/012016

Application of Natural Dolomite for Soil Upgrading

Jakapan Pimolrat, Kreangkrai Maneeintr*, Pinyo Meechumna

Carbon Capture, Storage and Utilization Research Group, Department of Mining and Petroleum Engineering, Faculty of Engineering, Chulalongkorn University, Bangkok 10330, Thailand.

*Krengkrai.M@chula.ac.th

Abstract. Soil degradation is the conditions with low-soil quality. It can result in low fertility, in soil. To overcome this problem, soil-quality upgrading should be applied. For this study, the acidic soil from Nan, Thailand is used to upgrade the soil conditions to fit well with the agricultural activities by using natural dolomite as a waste from the industries. It is blended with the acidic soil in various ratios from 5-30 % by weight. The optimal conditions for various soil properties such as pH, electrical conductivity (EC), bulk density and soil texture are investigated. From the results, the higher amount of dolomite can improve the soil properties. Acidity is lowered from 4.832 to 6.047. Bulk density is decreased at 1.1114 g/cm^3. Particle size distribution is sharply increased with the amount of dolomite. EC values are raised up to 4.25 dS/m. The natural dolomite can increase the soil quality from Nan province.

1. Introduction

Soil, the mixture of disintegrated rocks, organic matters, water, gas and some biological life forms, has been using for agricultural activities to serve foods for mankind [1]. However unwell management and improper usage of soil in the agriculture are greatly impact to its quality as can be affected as soil degradation which can decline crops productivity approximately 20% from 1998 - 2013 [2].

The soil degradation is definite that the decline of soil properties in term of physical, chemical and biological properties. In physical of soil, reducing pore structure as compaction, soil erosion on top layer soil and laterization on ironic soil are alarmed for decreasing of volume of water in the soil [3]. Lacking fertilizer is the major cause in chemical soil property along with soil pH as acidity and alkalinity and also the high rate leaching of cation [4]. The decline of biological property can be described in lowering rate of humus in soil. To reduce the soil degradation rate, the soil amendment should be applied to improve all soil conditions by adding some nutrient organic and inorganic matters [5].

Thailand, agricultural country, has been also facing the soil problems which pull down soil productivity on crops plant [6].The main land using type in Thailand is 53.69% for rural [7] and 26% of total area in Thailand requires treatment immediately to be maintained sustainable production [8].

In this study, soil from Nan, Thailand is collected from agricultural area. Its characteristic tends to be unacceptable in growing crops as it is moderately acidic, very silt texture, high bulk density. Upgrading the soil by blending natural dolomite in various ratios would regain suitable soil conditions for sustainable usage in agriculture.

Content from this work may be used under the terms of the Creative Commons Attribution 3.0 licence. Any further distribution of this work must maintain attribution to the author(s) and the title of the work, journal citation and DOI.
Published under licence by IOP Publishing Ltd

IOP Conf. Series: Materials Science and Engineering **859** (2020) 012016 doi:10.1088/1757-899X/859/1/012016

2. Methodology

2.1. Sample preparation
The acidic soil and natural dolomite were dried at 115 °C for 12 hours to remove moisture then the soil was screened by sieve no. 10 (2000 μm) to remove the coarse size. The passed sieve no. 10 soil was mixed by natural dolomite in 5, 10, 15, 20, 25 and 30 percentage by weight.

2.2. pH measurement
The soil, natural dolomite and blended samples were selected by 10 g Adding 25 ml distilled water and sample in the 50 ml beaker and stirred for 5 minutes then leave for 10 minutes for settling the samples.

Firstly the pH meter, Benchtop pH/Water Quality Analyzer LAQUA F-74, must be calibrated with buffer pH 7 and 4 respectively. After that all samples were measured for 3 times each.

2.3. *Electrical conductivity (EC) measurement*
The 10 g of dried soil, natural dolomite and blended samples were diluted with 50 ml distilled water into 100 ml beaker then stirred for 30 minutes and allowed the settling for 30 minutes. Afterward the settling samples were poured top layer liquid into the new beaker.

Calibrating Benchtop pH/Water Quality Analyzer LAQUA F-7, electrical conductivity with std. 0.01 N KCl then measured the samples for 3 times.

2.4. Specific gravity and particle density measurement
The 100 ml cylinder was filled by 50 ml distilled water. 20 g dried sample was added into the cylinder and the different of volume was measured. This different was replaced in formula as mentioned below.

$$\text{Specific gravity } (g/cm^3) = \frac{\text{Weight of sample (g)}}{\text{Water volume changed } (cm^3)} \tag{1}$$

More over the particle density could be calculated by using the above information and substituted into below formula.

$$\text{Particle density} = \frac{\text{Dry weight of soil (g)}}{\text{weight of soil + water (g)} - \text{weight of soil (g)}} \tag{2}$$

2.5. Bulk density measurement and soil porosity
The cylinder with diameter 5.94 cm and height 5.87 cm was used for this measurement. The 200 g dried samples were packed into the cylinder. The packable amount of each sample was taken out and measured its weight. The weight was calculated by following formula.

$$\text{Bulk density } (g/cm^3) = \frac{\text{oven dry weight of soil in cylinder (g)}}{\text{Volume of cylinder } (cm^3)} \tag{3}$$

$$\text{Soil porosity} = 1 - \frac{\text{bulk density } (g/cm^3)}{\text{sample's particle density } (g/cm^3)} \tag{4}$$

2.6. Particle size distribution analysis
40g Sodium hexametaphosphate 68% is diluted with 1,000 ml distilled water as a dispersion agent. The sample is placed into 500 ml beaker, add 125 ml dispersion agent and distilled water then soak it for 16 hours. Next, sample is poured to 1,000 ml cylinder then apply distilled water to reach 1,000 ml volume. Cover the cylinder with film and shake it for 1 min. The hydrometer in 0.995 – 1.038 in 0.001 divisions at 68 °F (20 °C) is used in order to measure the specific gravity of water, dropped gently into cylinder to detect the specific gravity at time 0, 1, 2, 5, 10, 15, 30, 60, 120, 250, 420 and 1,440 minutes, respectively. Meanwhile the temperature of water is measured in the same time.

For the next step, the sample is analyzed the coarse size by wet sieve in 10 (2,000 μm), 40 (400 μm), 80 (177 μm), 120 (125 μm) and 200 mesh (74 μm). All retained samples are dried in oven with

115 °C and measured the weight. All the obtained data has been replaced in the following formulas in order to find out size and cumulative percentage of passing of particle.

3. Results and discussion

According to the figure 1, the increasing of bulk density depends on among of natural dolomite blended in samples. There is 1.2119 g/cm^3 from original soil into 1.114 g/cm^3 on 30% dolomite which is requirement for ideal bulk density of plants growth [9]. It can represent that the higher volume of natural dolomite could raise slightly space for some water and gas in the soil. This evidence is also supported by higher volume of porosity in Figure 2.

In soil texture, 50% size distribution of particle (d50) in original soil is around 12. The trend of size distribution of particle seems to dramatically grow for high ratio of dolomite, there are 82, 94, 97, 110, 130 and 200 μm for 5, 10, 15, 20, 25 and 30% dolomite respectively. As the result, the increasing of grain size from silt loam to sandy loam can reflect the better drainage management in soil.

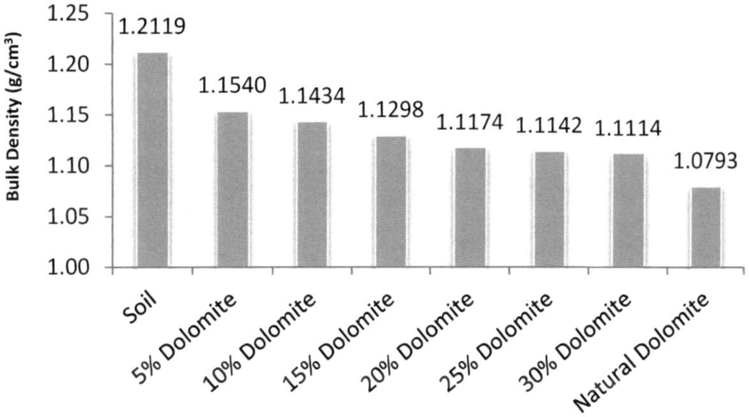

Figure 1. Bulk density of samples in percentage.

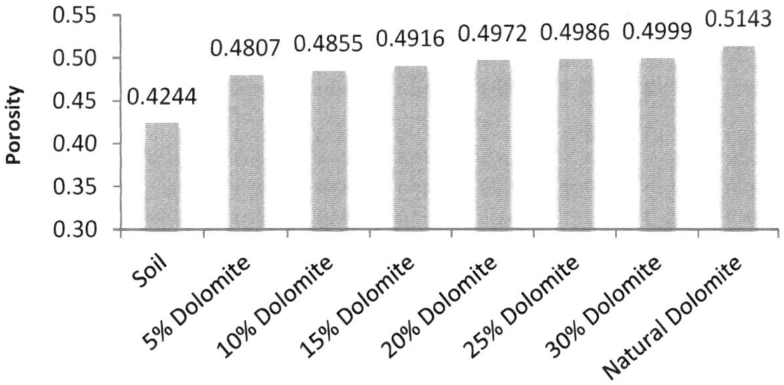

Figure 2. Porosity in samples.

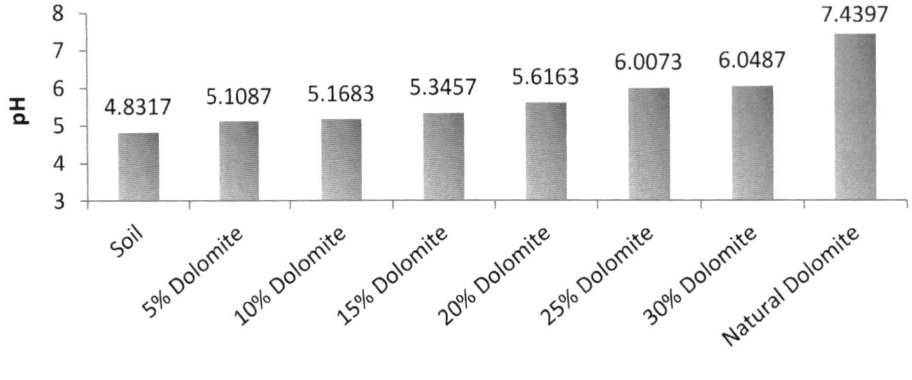

Figure 3. pH of samples.

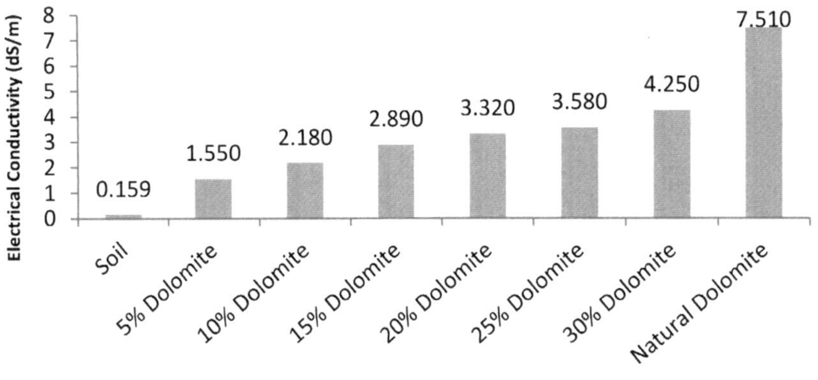

Figure 4. Electrical conductivity of samples.

As can be seen from figure 3, the acidity of original soil has been solved. The sample which has removed freely cation from uptaking and harvesting, resulted as acidic soil can be reduced acidity by dolomite. The chemical reaction of adding dolomite can create Ca^{2+} and Mg^{2+} to reach soil health standard. pH of samples has moved from pH 4 in acidic zone into 6.0487 for the highest ratio of dolomite. The blended sample is optimising suitable pH for crops growth in Thailand which is required pH 5.5-7.0 [10].

The electrical conductivity can represent to salinity of soil. All samples significantly increase EC especially 30% dolomite which is 4.25 dS/m as shown in figure 4. However, the significantly increasing of EC seems to damage the plants growth. Over 2.00 dS/m EC can reduce potato crops yield for 20% [11] and also over 7.00 dS/m extremely affects for all crops growth [12].

4. Conclusion

To upgrading the acidic soil from Nan, Thailand, blending natural dolomite can resolve the soil degradation in term of both physical and chemical properties. The high among of dolomite blending into the soil could recover the soil to gain capacity of air and water containing from higher porosity and improving soil aggregation, pH problem moreover has been adjusted for suitable with many type of crops. However raising ratio of dolomite can turn the soil to be salty which may need other soil amendments combined to reach the best solution.

Acknowledgments
The authors would like to acknowledge the Malaysia-Thailand Joint Authority (MTJA) for financial support of this project.

References
[1] Richard Y, Stefano O and Ian F 2015 *Soil Degradation:A Major Threat to Humanity* 1
[2] United Nations Development Programme 2019 *Combatting Land Degradation Securing A Sustainable Future* 1
[3] Pagliai M and Vignozzi N 2003 *The Soil Pore System as an Indicator of Soil Quality* Available from: https://www.semanticscholar.org/paper/The-Soil-Pore-System-as-an-Indicator-of-Soil-Pagliai-Vignozzi/81a867b7dbe23b8c08eb7d778b3515a5377a364c
[4] McCauley A, Jones C and Olson-Rutz K 2017 *Nutrient Management Module 8* 1
[5] Clements D P and Bihn E A 2019 *Safety and Practice for Organic Foods* 321
[6] Tonmanee T and Kanchanakool N 1999 *Water Science & Tachnology* 39 61
[7] Land Development Department of Thailand Land 2012 *Use Types of Thailand in 2009-2012* Available from: http://www.ldd.go.th/ldd_en/en-US/land-use-types-of-thailand-in-2009-2012/
[8] Moncharoen P, Vearasilp T and Eswaran H 2001 *Sustainable for Global Farm* 179
[9] Cliffie W, Sagona J, Kachala O, Matete S and Jenya H 2016 *Universal Journal of Agricultural Research* 4 155
[10] Food and Agriculture Organization of the United Nations 2017 *Final Report National Agro-Economic Zoning for Major Crops in Thailand (NAEZ) (Project TCP/THA/3403) NAEZ Model Implementation and Results* 41
[11] Reis Jr R A, Fontes P C R, Neves J C L and Santos N T 1999 *Science Agricola* 56 993
[12] Maas E V and Hoffman G J 1997 *Journal of the Irrigation and Drainage Division* 103 115

IOP Conf. Series: Materials Science and Engineering **859** (2020) 012017 doi:10.1088/1757-899X/859/1/012017

Performance Measurement and Improvement of Lean Manufacturing Operations: A Leanness Assessment Literature Review for the Product Development Industry

Leandro Silvério, Luís Gonzaga Trabasso, Marcus Vinicius Pereira Pessôa

Aeronautics Institute of Technology (ITA), Praça Marechal Eduardo Gomes, 50—Vila das Acácias, 12228–900 São José dos Campos/SP – Brazil

leandro.silverio.pereira@gmail.com

Abstract. Today manufacturing sectors are more competitive than before. Thus, to execute an enterprise's transformation in a company emerging as a lean organization it is crucial to have assessments and performance measurements that observe multiple variables during the lean implementation patch. The objective of this paper is to demonstrate a research method focused on lean performance measurement assessments from previous works, summarizing and organizing the existing evaluation standards in a way to allow different industrial segments to replicate the screening steps for a literature review construction. As a result, the existing lean performance measurements and methods from the last 23 years were refined in order to highlight opportunities and insights for future researches to be developed on the lean field.

1. Introduction

A good manufacturing practice for production assembly can result in better productivity for the whole organization [1]. In order to improve the manufacturing processes of a firm, lean manufacturing (LM) is being adopted by enterprises since the early phases of a project; in a way to allow managers and team leaders to rapidly implement changes on product characteristics or requirements definition [2]. Lean manufacturing also relies on a series of tools and techniques focused on the manufacturing operations, including the product development (PD) industry. These are based on concepts from different applied sciences; likewise, a number of models are implemented for production process optimization to improve its characteristics or attributes from integral and differential calculus, accelerated approach methods, and others. Such applications are found in reordering inventory management points as well as in deterministic and stochastic operation researches, where uncertainties and risks are integrated into the estimates. In other words, lean manufacturing applies a wide variety of engineering and applied science techniques to support organizations on identifying the bottlenecks for the main processes [3]. Many authors have tried to describe the lean ever since it first appeared in the literature [4-11]. In common to all of them are waste elimination and the focus on available resource optimization [12]. According to Morgan and Liker, a lean company (LC) aims for lean practices along the whole lean development process (LDP), not only for the manufacturing shop phases [13,14] and working as a single–piece manufacturing flow [15]. Various methods for an enterprise's operational performance transformation are comprised under the lean strategy umbrella [16] and researchers have already demonstrated that implementing lean along a product development process (PDP), can support an enterprise to emerge as a lean organization and increase its overall operational results [17-21].

Content from this work may be used under the terms of the Creative Commons Attribution 3.0 licence. Any further distribution of this work must maintain attribution to the author(s) and the title of the work, journal citation and DOI.
Published under licence by IOP Publishing Ltd

2. Background

Each company's sector has expertise on its own development phases, that way when the process is in progress significant data can be collected [22]. Thus, the system solution shall be considered as a major input for the decision–making process by assessing every development phase, allowing companies and managers to react to changes with countermeasures [23] by attacking the leanness level performance results early in the project, supported by all stakeholders [24]. A leanness level is defined by Vinodh and Chintha [25] as a performance measure of lean operations. Comm and Mathaisel [26] and Bauch [27] also described leanness as a relative measure of whether a company is lean or not. The number of studies in the literature on leanness assessment is low compared to those on lean implementation areas [28] and according to Silvério et al. [29], there is also a lack of self–assessment methods for the leanness definition, associated with a roadmap to provide insights for decision makers.

3. Methodology

In order to provide insights for future researches on performance measurement and improvement of lean manufacturing operations, a literature review and results classification were performed according to the criteria presented in the work from Silvério et al. [29], where the collected data was synthesized and deployed as per the flowchart presented in the Figure 1. Same flowchart can also be replicated for different industrial segments, serving as a support tool for a literature review construction.

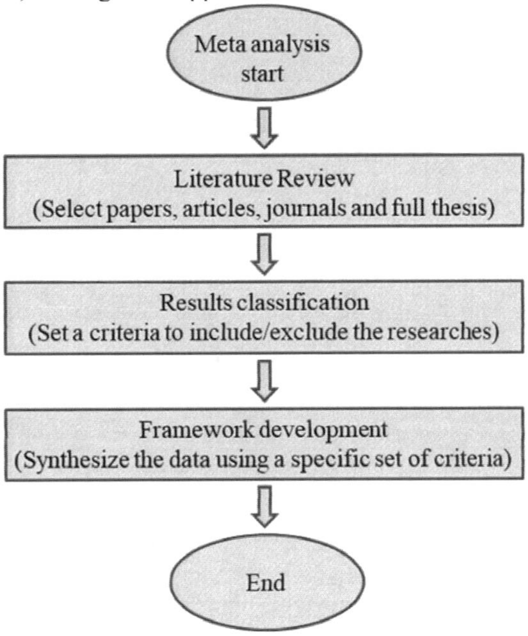

Figure 1. Research flowchart method (constructed by authors).

Similar to other literature review papers, the only restriction was that the reviewed articles were in English [30]. From the assessed studies, over 90 papers, articles, journals, and full thesis from the last 23 years were screened and refined in order to compose a systematic literature review and results classification according to the following criteria, presented in the Table 1 in a chronological sequence.

A) Is the model approach a qualitative or quantitative assessment?

B) Does the model have defined indicators?

C) Does the model provide the enterprise's leanness level?

D) Does the model provide a lean roadmap after the assessment is performed?

Table1. Classification of literature sources framework. [29]

Author	A	B	C	D	Author	A	B	C	D
Karlsson and Åhlström	L	Y	N	N	Wong and Lai	T	Y	Y	N
Detty and Yingling	LT	Y	Y	Y	Kuhlang et al.	T	Y	Y	Y
Sánchez and Pérez	LT	Y	Y	N	Eroglu and Hofer	LT	Y	Y	N
Soriano-Meier and Forrester	T	N	Y	N	Chauhan and Singh	L	Y	Y	N
Nightingale and Mize	LT	Y	Y	Y	Vinodh and Vimal	L	Y	Y	Y
Goodson	LT	Y	Y	N	Nasab et al.	L	N	Y	N
Kumar and Thomas	LT	Y	Y	N	Azevedo et al.	L	Y	Y	N
Hon	LT	Y	Y	N	Anvari et al.	T	Y	Y	Y
Shah and Ward	LT	Y	N	N	Bhasin	L	Y	Y	N
Leung and Lee	L	Y	Y	N	Amin and Karim	T	Y	N	Y
Hobbs	LT	Y	N	Y	Karim and Arif-Uz-Zaman	T	Y	Y	Y
Kojima and Kaplinsky	LT	Y	Y	N	Gupta et al.	T	Y	Y	Y
Doolen and Hacker	LT	Y	Y	N	Alemi and Akram	T	Y	Y	N
Little and McKinna	LT	Y	Y	Y	Behrouzi and Wong	T	Y	Y	Y
Taj	LT	Y	Y	N	Mostafa et al.	LT	Y	Y	N
Wan and Chen	T	Y	Y	N	Wahab et al.	L	Y	Y	N
Ray et al.	T	Y	Y	N	Lemieux et al.	LT	Y	Y	Y
Wan	LT	Y	Y	N	Al-Najem et al.	LT	Y	Y	Y
Bonavia and Marin	LT	Y	Y	N	Al–Ashaab et al.	LT	Y	N	Y
Srinivasaraghavan and Allada	T	Y	Y	Y	Lucato et al.	LT	Y	Y	N
Wan et al.	T	Y	Y	Y	Elnadi and Shehab	LT	Y	Y	N
Shah and Ward	LT	Y	Y	N	Pakdil and Leonard	LT	Y	Y	Y
Matsui	LT	Y	N	N	Nesensohn et al.	L	Y	Y	N
Sanati and Seyedhoseini	T	Y	Y	N	Ramirez and Lorena	LT	Y	Y	Y
Dal Pont et al.	LT	Y	N	Y	Hosseini and Ebrahimi	L	Y	Y	Y
Barad and Dror	L	Y	N	Y	Mostafa et al.	LT	Y	N	Y
Bayou and de Korvin	LT	Y	Y	N	Soltan and Mostafa	LT	Y	Y	N
Bhasin	L	Y	Y	Y	Donovan	LT	Y	Y	N
Saurin and Ferreira	L	Y	Y	N	Urban	T	Y	Y	N
McLeod	LT	Y	N	N	Mahfouz and Arisha	LT	Y	Y	Y
Gurumurthy and Kodali	LT	Y	Y	Y	Vidyadhar et al.	LT	Y	Y	Y
Wu and Wee	L	Y	N	N	Omogbai and Salonitis	LT	Y	Y	Y
Marvel and Standridge	LT	Y	Y	Y	Maasouman and Demirli	LT	Y	Y	Y
Puvanasvaran et al.	L	Y	Y	N	Carvalhosa et al.	LT	Y	Y	N
Rahman et al.	LT	Y	N	N	Leite et al.	LT	Y	N	Y
Jeyaraman and Teo	L	Y	Y	Y	Hjalmarsson and Olsson	LT	Y	Y	Y
Singh et al.	LT	NA	Y	N	Abreu and Calado	LT	Y	Y	Y
Zanjirchi et al.	LT	NA	Y	N	Rajpurohit et al.	LT	Y	Y	Y
Sun	LT	NA	Y	N	Galankashi and Helmi	LT	Y	Y	N
Nordin et al.	L	Y	Y	N	Gonçalves and Salonitis	LT	Y	N	N
Anvari et al.	L	Y	N	Y	Sangwa and Sangwan	LT	Y	Y	N
Asadi and Panahi	T	Y	Y	N	Albzeirat et al.	LT	Y	Y	N
Aurelio et al.	LT	Y	Y	Y	Bento and Tontini	LT	Y	Y	N
Anvari et al.	LT	Y	N	Y	Rakhmanhuda and Karningsih	LT	Y	Y	Y
Bhasin	LT	Y	Y	N	Belhadi et al.	LT	Y	Y	N
Seyedhosseini et al.	L	Y	Y	N	Pakdil et al.	L	Y	Y	Y
Vinodh and Chintha	LT	Y	Y	Y	Aikhuele and Turan	LT	Y	Y	N
Vinodh and Balaji	LT	Y	Y	Y	Silvério et al.	L	Y	Y	Y

Y=Yes; N=No; NA=Not Applicable; L=Qualitative; T=Quantitative; LT=Qualitative and Quantitative.

4. Results and analysis

Karlsson and Ahström [31] presented the first lean assessment model found in the literature in 1996. It consisted of a checklist to determine if a company is adopting lean aspects on its lean journey or not through a set of nine measurable determinants on waste elimination. The model had the purpose of evaluating the manufacturing progress of an enterprise focusing on lean production implementation, and it is based on the lean core principles described in the book "The Machine that Changed the World" by Womack, Jones, and Roos [19], where the term LM was coined by Womack after his visit to Japan, to witness the effectiveness of the Toyota Production System (TPS) [32]. Although the first leanness research was published in 1996, the next leanness study appeared in the literature only four years later in 2000, and was by Detty and Yingling [33]. The authors published a case study focused on quantifying the lean conversion benefits in a manufacturing enterprise. The framework provided recommendations for an existing manufacturing enterprise to become lean based on a discrete event simulation tool supporting managers' decisions on implementing leanness or not. The operational benefits quantification of each lean principle was considered by the method.

From 1996 on, multiple leanness assessment methods, frameworks and LM concepts were published by different authors over the years; including the work from Silvério et al. [29] that was used as results classification and criteria to include/exclude the researches for the literature review analysis and conclusions presented by this study.

As a result, this paper compiled various performance measurement and improvement of LM operations in the PD industry. A bibliographic research was conducted to better organize the research problem [34] and the proposed framework from Silvério et al. [29] was used as results classification and criteria to include/exclude the researches for the literature, dividing the selected papers into four categories: qualitative or quantitative assessment model, indicators defined or not, leanness level defined or not and lean roadmap provided or not. As a result, there were (21) qualitative, (16) quantitative and (59) qualitative/quantitative researches among the literature sources. In addition, (91) papers defined indicators and (2) not. For the leanness level definition, (80) researches proposed a leanness level classification and (16) not. Regarding the roadmap being provided after the assessment performed, (42) papers presented a roadmap and (54) not. Thus, from the screened researches it was identified a reduced number of papers focused on lean implementation roadmaps; and a widely range of practical studies on LM implementation practices. In order to reinforce an unifying view of the leanness and roadmap practices concept, we propose the following definition of leanness: 'Leanness is an overall progress measurement of the lean production in a firm, focused to determinate where the assessed enterprise successfully applies the lean initiatives and where the lean implementation practices must be incorporated or improved, substantiated by successful application cases and models.'

5. Conclusions

As a conclusion, leanness methods are in a wide range, varying from qualitative/quantitative checklists to multiple complex numerical models. Furthermore, the leanness research field varies and continues to be analyzed in order to possibly compose a standardized process for enterprises emerging as true lean organizations. Thus, the analytical hierarchy process (AHP) can also serve as a reference methodology for future studies to be developed on the leanness performance measurement and improvement of LM operations, as far as by applying AHP in the selected determinants the fundamental aspects could be accurately analyzed in a multi–factor manner [35]. This paper will also serve as insights for future researches and practitioners focused on the new performance measures, tools and methods for the lean operations; and as a practical guidance do be considered by managers and decision makers when selecting a leanness framework to be deployed on their organizations. Moreover, this paper will serve as a screening reference method for conducting a literature review in different industrial segments, as far as it demonstrated to be a useful support tool to organize the existing evaluation standards in a way to allow different industrial segments to replicate the screening steps for a literature review construction, as demonstrated by Figure 1 research flowchart method and Table 1 classification of literature sources framework.

Acknowledgments
The author(s) declared no potential conflicts of interest with respect to the research, authorship, and/or publication of this article. The author(s) received no financial support for the research, authorship, and/or publication of this article.

References
[1] Fansuri AFH, Rose ANM, Ab. Rashid MFF, Mohamed Nik NMZ and Ahmad H (2018) Productivity Improvement Through Line Balancing at Electronic Company – Case Study. In: *International Conference on Manufacturing, Material and Metallurgical Engineering (ICMMME).* Kuala Lumpur, Malaysia, 17–19 March 2018. IOP Publishing. DOI: doi.org/10.1088/1757-899X/409/1/012015.
[2] Martini LGS and Trabasso LG (2005) Product Strategic Development (PSD) in the Aviation Segment. In: *18th International Congress of Mechanical Engineering*. Ouro Preto, Minas Gerais, Brazil, 06–11 November 2005. ABCM.
[3] Cordeiro E, Barbosa GF and Trabasso LG (2016) A customized QFD (quality function deployment) applied to management of automation projects. *The International Journal of Advanced Manufacturing Technology*. DOI: doi.org/10.1007/s00170-016-8626-0.
[4] Hines P, Holweg M and Rich N (2004) Learning to Evolve: A Review of Contemporary Lean Thinking. *International Journal of Operations & Production Management* 24(10): 994–1011. DOI: doi.org/10.1108/01443570410558049.
[5] Hines P, Francis M and Found P (2006) Towards lean product lifecycle management: A framework for new product development. *Journal of Manufacturing Technology Management* 17(7): 866–887. DOI: doi.org/10.1108/17410380610688214.
[6] Lewis MA (2000) Lean production and sustainable competitive advantage. *International Journal of Operations & Production Management* 20(8): 959–978. DOI: doi.org/10.1108/01443570010332971.
[7] Possamai O and Ceryno PS (2008) Lean approach applied to product development. *Product: Management & Development* 6(2): 157–165.
[8] Shah R and Ward PT (2007) Defining and developing measures of lean production. *Journal of Operations Management* 25(4): 785–805. DOI: doi.org/10.1016/j.jom.2007.01.019.
[9] Karningsih P, Pangesti AT and Suef M (2019) Lean Assessment Matrix: A Proposed Supporting Tool for Lean Manufacturing Implementation. In: *Annual Conference on Industrial and System Engineering (ACISE)*. Semarang, Central Java, Indonesia, 23–24 April 2019. IOP Publishing. DOI: doi.org/10.1088/1757-899X/598/1/012082.
[10] Rojas–Suarez JP, Gómez–Colmenares C and Vergel–Ortega M (2019) Lean six sigma in academic processes. In: *V International Conference Days of Applied Mathematics*. Barranquilla, Colombia, 15–17 May 2019. IOP Publishing. DOI: doi.org/10.1088/1742-6596/1414/1/012006.
[11] Gao T and Gurd B (2019) Organizational issues for the lean success in China: exploring a change strategy for lean success. *BMC Health Services Research* 19. DOI: doi.org/10.1186/s12913-019-3907-6.
[12] Karam A–A, Liviu M, Cristina V, et al. (2018) The contribution of lean manufacturing tools to changeover time decrease in the pharmaceutical industry. A SMED project. *Procedia Manufacturing* 22: 886–892. DOI: doi.org/10.1016/j.promfg.2018.03.125.
[13] Morgan J and Liker J (2006) *The Toyota Product Development System.* New York, NY: Productivity Press. DOI: doi.org/10.4324/9781482293746.
[14] Pessôa MVP and Seering W (2014) Trapped on the waste net: a method for identifying and prioritizing the causes of a corporations low product development performance. In: *Proceedings of the 13th international design conference*. Dubrovnik, Croatia, 19–22 May 2014. Ed DESIGN.
[15] Bonnal P, De Jonghe J and Ferguson J (2006) A deliverable–oriented EVM system suited to a large–scale project. *Project Management Journal* 37(1): 67–80. DOI: doi.org/10.1177/875697280603700106.

[16] Bhasin S (2012) Performance of lean in large organisations. *Journal of Manufacturing Systems* 31(3): 349–357. DOI: doi.org/10.1016/j.jmsy.2012.04.002.
[17] Krafcik JF (1988) Triumph of the lean production system. *MIT Sloan Management Review* 30(1): 41–52.
[18] Ōno T (1988) *Toyota Production System: Beyond Large–Scale Production.* Boca Raton, FL: CRC Press.
[19] Womack JP, Jones DT and Roos D (1991) *The Machine that Changed the World.* New York: Harper Perennial.
[20] Hernandez–Matias JC, Ocampo J, Hidalgo A and Vizan A (2019) Lean manufacturing and operational performance: Interrelationships between human–related lean practices. *Journal of Manufacturing Technology Management.* DOI: doi.org/10.1108/JMTM-04-2019-0140.
[21] Panwar A, Jain R, Rathore APS, Nepal B and Lyons AC (2018) The impact of lean practices on operational performance – an empirical investigation of Indian process industries. *Production Planning & Control* 29. 158–169. DOI: doi.org/10.1080/09537287.2017.1397788.
[22] Negroni Y and Trabasso LG (2009) A Quality Improving Method to Assist the Integrated Product Development Process. In: *17th International Conference on Engineering Design.* California, United States, 24–27 August 2009. ICED'09.
[23] Borba JCR, Trabasso LG and Pessôa MVP (2019) Agile Management in Product Development. *Research–Technology Management.* DOI: doi.org/10.1080/08956308.2019.1638488.
[24] Branco M, Loureiro G and Trabasso LG (2011) Balancing Value in Complex Systems Development. *Advanced Concurrent Engineering.* DOI: doi.org/10.1007/978-0-85729-799-0_37.
[25] Vinodh S and Chintha SK (2011) Leanness assessment using multi–grade fuzzy approach. *International Journal of Production Research* 49(2): 431–445. DOI: doi.org/10.1080/00207540903471494.
[26] Comm CL and Mathaisel DFX (2005) A case study in applying lean sustainability concepts to universities. *International Journal of Sustainability in Higher Education* 6(2): 134–146. DOI: doi.org/10.1108/14676370510589855.
[27] Bauch C (2004) *Lean product development: making waste transparent.* Diploma Thesis, Massachusetts Institute of Technology, Cambridge. DOI: doi.org/10.1108/14676370510589855.
[28] Narayanamurthy G and Gurumurthy A (2016) Leanness assessment: a literature review. *International Journal of Operations & Production Management* 36(10): 1115–1160. DOI: doi.org/10.1108/IJOPM-01-2015-0003.
[29] Silvério L, Trabasso LG and Pessôa MVP (2019) A Roadmap for a Leanness Company to Emerge as a True Lean Organization. *Concurrent Engineering: Research and Applications.* 28(1), 3–19. DOI: doi.org/10.1177/1063293X19888259.
[30] Pessôa MVP and Jauregui-Becker J (2020) Smart design engineering: a literature review of the impact of the 4th industrial revolution on product design and development. *Research in Engineering Design.* 175–195. DOI: doi.org/10.1007/s00163-020-00330-z.
[31] Karlsson C and Åhlström P (1996) Assessing changes towards lean production. *International Journal of Operations & Production Management* 16(2): 24–41. DOI: doi.org/10.1108/01443579610109820.
[32] Khusaini N, Ismail A and Amirul A (2016) Investigation of the prominent barriers to lean manufacturing implementation in Malaysian food and beverages industry using Rasch Model. In: *2nd International Manufacturing Engineering Conference and 3rd Asia-Pacific Conference on Manufacturing Systems (iMEC-APCOMS).* Kuala Lumpur, Malaysia, 12–14 November 2015. IOP Publishing. DOI: doi.org/10.1088/1757-899X/114/1/012090.
[33] Detty RB and Yingling JC (2000) Quantifying benefits of conversion to lean manufacturing with discrete event simulation: a case study. *International Journal of Production Research* 38(2): 429–445. DOI: doi.org/10.1080/002075400189509.

IOP Conf. Series: Materials Science and Engineering **859** (2020) 012017 doi:10.1088/1757-899X/859/1/012017

[34] Rocha G, Alfredson J, Trabasso LG, Almeida A, Matheus A, Arjoni D and Rodriguez Diaz M (2018) A Roadmap for Assessing Human Factors Influence on Pilot Performance Based on Simulates Flights. In: Conference: 31st Congress of the International Council of the Aeronautical Sciences. Belo Horizonte, Brazil, 09–14 September 2018. ICAS.

[35] E Jayamani, DS Perera, KH Soon and MKB Bakril (2017) Application of Analytic Hierarchy Process (AHP) in the analysis of the fuel efficiency in the automobile industry with the utilization of Natural Fiber Polymer Composites (NFPC). In: *International Conference on Manufacturing, Material and Metallurgical Engineering (ICMMME)*. Bangkok, Thailand, 17–19 March 2017. IOP Publishing. DOI: doi.org/10.1088/1757-899X/191/1/012004.

Developing a Trajectory Planning for Curved-Contoured Surfaces for Use by 8-DoF Workcell in Robotic Fibre Placement

Mohamed M.A. Ammar[1, *], Bijan Shirinzadeh[1], Pan Zhao[2], Yaoyao Shi[2]

[1] Robotics and Mechatronics Research Laboratory, Department of Mechanical and Aerospace Engineering, Monash University, Clayton 3800, Australia
[2] The Key Laboratory of Contemporary Design and Integrated Manufacturing Technology, Ministry of Education, Northwestern Polytechnical University, Xi'an, Shaanxi 710072, China

mohamed.ammar@monash.edu

Abstract. The replacement rate of conventional material by composites is increasing. Possessing high specific strength and stiffness makes composites attractive to many applications. Different techniques are being utilized for composite placement processes. Robotic Fibre Placement (RFP) is introduced as a competing approach for composite fabrication. This approach can provide many advantages; e.g. low labor cost, high performance, quick and effective process. The current work proposes an 8-DoF system established for laying the composite material. This system composed of three main units includes a 6-DoF industrial robot and two mandrel tools with different configurations. A sensory-based feedback control system is developed to manage the placement process. The placement process can be performed on two different surfaces. The end-cap surface has been fabricated then the cylindrical surface. Different geometries of pressure vessels have been generated; e.g. Oxidizer tanks that can be used in many applications such as (Hybrid rockets).

1. Introduction

Recently, composites have gained popularity as an alternative material in many industrial sectors because of their excellent properties [1-4]. Fibrous composites are the mostly used composites in many applications, especially in the aerospace industry. Carbon fibre reinforced polymers (CFRP) became one of the most significant fibrous composites after the growth of applications and experience in this field because of their superior performance [5, 6].

Composite lay-up is one of the most important challenges facing the wider use of composite products. There are many traditional techniques of manufacturing composite structures such as hand lay-up and tape laying [7, 8]. Those techniques are using a manual placement of both fibre and matrix materials on a fabricated mold with the required geometry. They have many drawbacks such as time-consuming, hazardous, labor-intensive, and low efficiency, which result in a high material scrap rate [9]. Therefore, the conventional methods may be combined with the computer-controlled systems, whereas online consolidation was automated carefully to obtain accurate products. Accordingly, three techniques of automated consolidation are being used, Automated Tape Laying (ATL), Filament Winding (FW) and Robotic Fibre Placement (RFP) [10, 11]. RFP is introduced as innovative techniques that can overcome most drawbacks of conventional techniques, as it can satisfy all current in-

Content from this work may be used under the terms of the Creative Commons Attribution 3.0 licence. Any further distribution of this work must maintain attribution to the author(s) and the title of the work, journal citation and DOI.
Published under licence by IOP Publishing Ltd

dustry requirements. Applying this procedure in the industry can provide cost-effective, high flexibility, faster and highly accurate process. Hence, it can be used for all intricate structures [9, 12-15].

Many research works have been done on different techniques of fibre placement. Sofi et al. [16] established a review of the dry fibre winding processes. Different winding techniques were explored according to the application geometry, e.g. geodesic and non-geodesic formulations. The review introduced some innovative strategies for fibre path planning in the upcoming research work. However, the authors pointed out there still are some challenges to apply their strategies on complex shapes. Rousseau et al. [17] outlined another review about the multiple lay-up techniques that were performed for Automated Fibre Placement (AFP) processes. Different lay-up strategies were described using the numerical and mathematical work of the fibre placement on the conical, cylindrical, curved, planer and perforated planer surfaces.

Shirinzadeh et al. [9] outlined the procedure of the RFP process. Three different methods were described, including fibre path generation based on simulation, fibre steering, and sensory-based contour. A straight line and a circular arc trajectory were used for the fibre placement process on flat and complex surfaces. Their work has been extended on RFP process in [18]. An automatic fibre path was generated using an innovative algorithm, which was applied for opened and closed surfaces. Another path generation on open-contoured structures using a surface curve algorithm (SCAR), has been shown in [19]. The proposed methods were conducted to produce a set of curved and flat surfaces to ensure that the RFP paths of the lamina were smooth without any drawbacks. Another path planning technique for the compaction roller on different geometries has been produced in [13]. The proposed technique was used for free-form shaped structures. The path planning research has been continued for more complex structures. For example, Hely et al. [20] used a robotic fibre placement system to manufacture Y-shaped tubes. Hassan et al. [21] performed another fibre placement process for a dome-shaped surface, using multiple Autonomous Industrial Robots (AIRs). Zhao et al. [22] also presented an approach for path planning but for the aircraft tail using AFP.

Based on the review, most of the path planning work was for the open contoured surfaces. No experimental work has been done for manufacturing pressure vessels using RFP, especially the closed surfaces with extreme curvatures. Therefore, the present work proposes a novel technique which aims to use RFP to fabricate the cylindrical pressure vessels, e.g. the oxidizer tank which can be included in many implementations such as (Hybrid rockets). An 8-DOF workcell is established to perform the placement process. It is including an industrial robot which is connected to two, horizontal and vertical, mandrel tools. Path planning algorithm is constructed for the proposed surfaces. Also, a force/torque feedback system is utilized to control the path generation process. The following sections will outline the proposed process in-details.

2. Sensory based lay-up

The direction of fibre placement plays an important role in calculating the final mechanical properties of the composite products. To obtain the accurate lay-up angles, a suitable trajectory for the robot end-effector should be generated. The current work is proposing a path generation method that uses a sensory-based feedback control system. The robot trajectories are produced based on a real-time process. This method utilizes the information of sensory-based feedback signals to define the tag points which are generated to define the placement trajectory. A 6-DoF force/ torque sensor is installed in the robot head to monitor the compaction force for the placement process. The compaction force is a vital aspect for ensuring the best consolidation of the fibre tows on the substrate surfaces. The end-effector trajectory is corrected using the collected data to maintain the same value of the compaction force between the robot head and the workpiece surface. This force is also adjusted to be normal on the workpiece surface during the whole process based on the proposed algorithm in [9]. This technique doesn't need any mathematical model as the end-effector direction is adjusted to be always normal to the structure surface. The data of the force/torque sensor is compared with the determined values to calculate the exact angle between the end-effector and the surface normal, based on the relationship developed in

[9]. Finally, after generating the correct paths, application programs are written to define all placement paths. Then the generated programs are uploaded to the robot controller using the robot software.

3. Experimental setup

The lay-up procedure is performed based on a synchronized motion between the robot end-effector and the workpiece surface. We established an innovative workcell which composes of three main units to obtain the accurate motion. The first unit is a fibre placement head which is operated with a 6-DoF industrial robot (Yaskawa Motoman SK120) as shown in Fig. 1. Prepreg tows for the fibrous composite have been used in this work. The tow motion occurs throw the creel rollers system until reaching the compaction roller.

Fig. 2a. presents the second unit (Vertical mandrel tool) which is used to manufacture the end-cap surface of the cylindrical pressure vessel. A cylindrical tool is moved by a servo-motor using a gear system. The motor motion is controlled by feedback signals using a PCI controller card. The motor position, velocity, and acceleration are controlled based on the end-effector motion. Fig. 2b. shows the third unit (Horizontal mandrel tool). The presented system is used to perform fibre placement on the circumferential surface of the pressure vessel. The lay-up on that surface is applied with different angles, that will be described in Section 4. The servo motor is also controlled using feedback signals from the motor encoder to the robot main control unit.

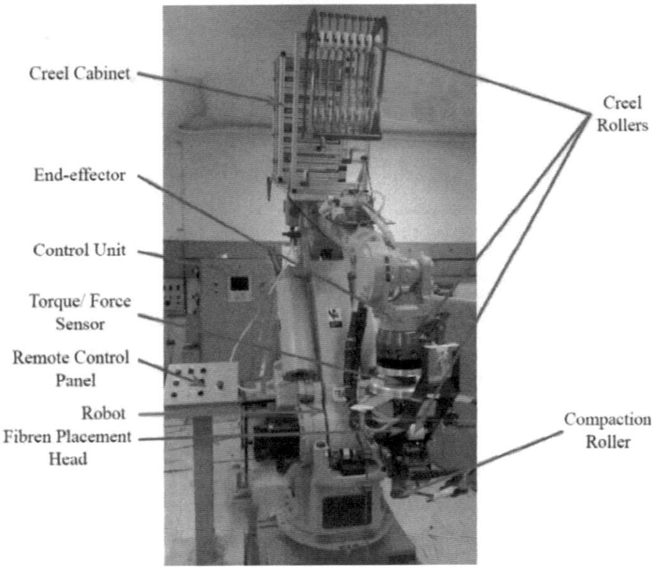

Figure 1. Robotic fibre placement facility (SK120).

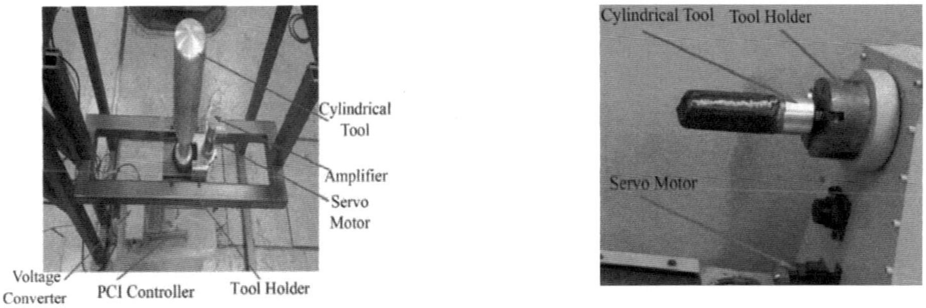

(a) Vertical mandrel tool. (b) Horizontal mandrel tool.

Figure 2. Mandrel tools.

IOP Conf. Series: Materials Science and Engineering **859** (2020) 012018 doi:10.1088/1757-899X/859/1/012018

4. Results and discussion

4.1. Generated trajectories

Offline programming is used to manage the workcell operation in the placement process. Robot application programs are created based on the required path directions. An operation algorithm is used for each required placement angle as presented in Fig. 3. Therefore, the fibre placement on the cylindrical pressure vessel surface has been obtained based on two stages. Firstly, the circumferential surface of the pressure vessel has been fabricated using four different trajectories as presented in Fig. 4. Fig. 4a. shows fibre placement on 0° angle. The robot end-effector is moved in a longitudinal direction to the endpoint of the path. Subsequently, during return time, the mandrel tool is rotated with a defined angle according to the width of prepreg tows. Fig_s. 4b. and 4c depict the paths of 45° and -45° angles, respectively. The end-effector is moved in an inclined angle with the slow rotation of the cylindrical tool to produse the required paths. Finally, Fig. 4d. shows the trajectory obtained of 90° angle. The end-effector is moved to the start point, then the mandrel tool revolved continuously until the end of the process.

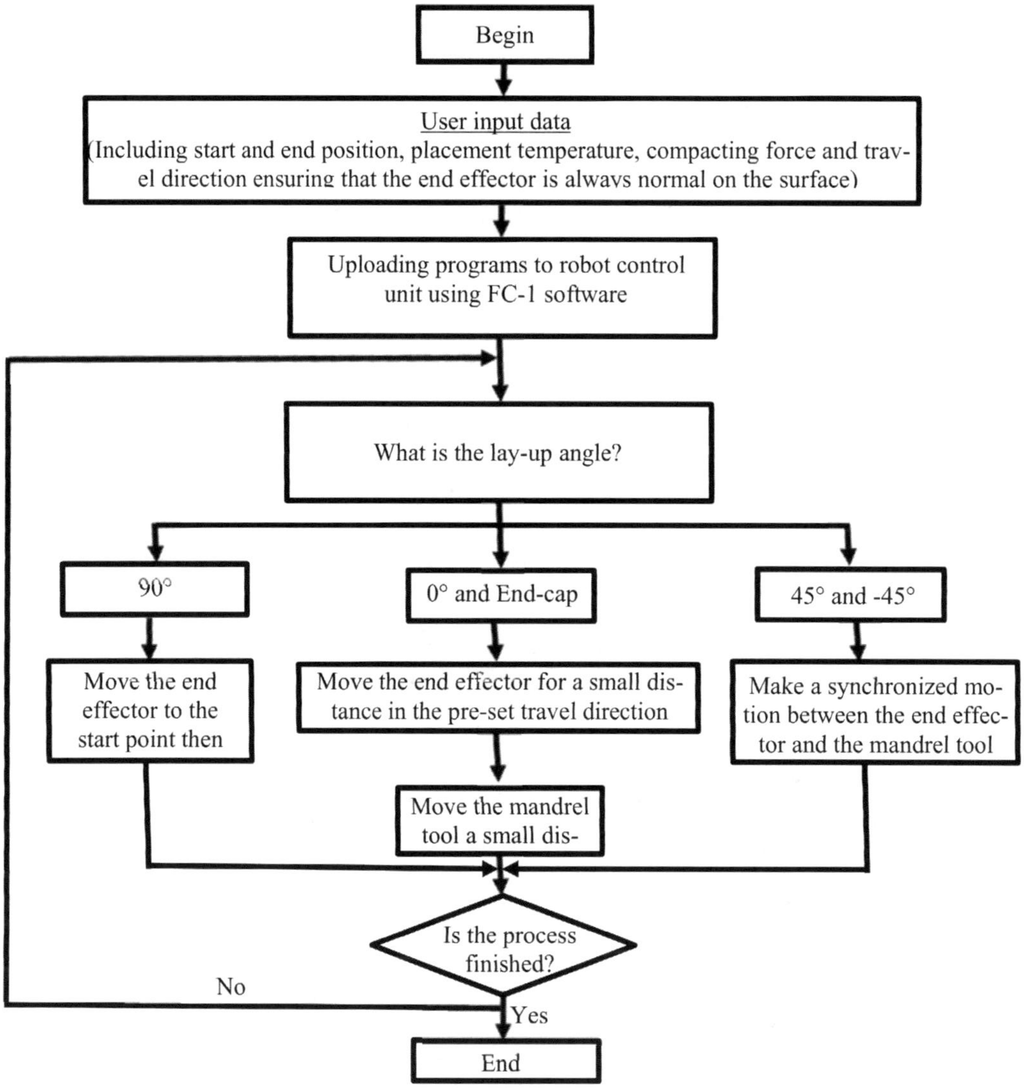

Figure 3. Path generation algorithm.

Secondly, the end-cap part is fabricated based on one predefined trajectory as shown in Fig. 5. The robot head is moved in a series of linear and spline motions until finishing the first increment. Subsequently, the mandrel tool revolved using step motion during the return time of the end-effector.

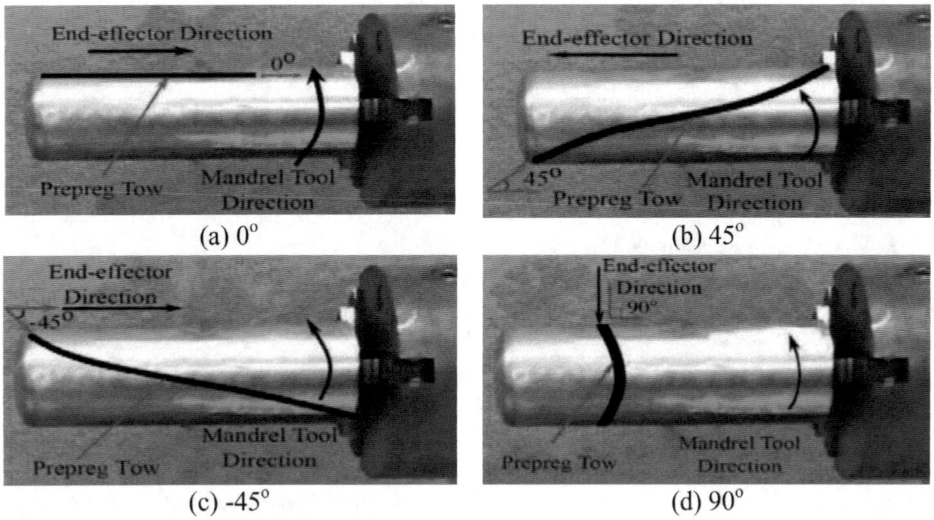

Figure 4. Circumferential surface placement.

Figure 5. End-cap placement.

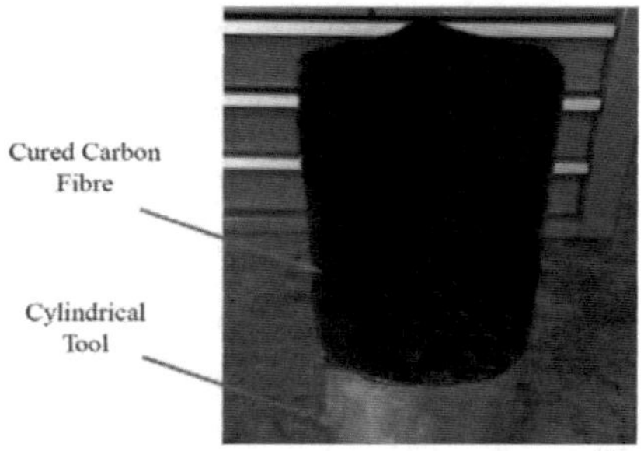

Figure 6. Circumferential surface placement.

4.2. RFP products

The proposed operation is applied using thermosetting composite prepreg tows. A uni-direction carbon fibre prepreg tows (T700G) from Toray America, Inc is used. A lay-up of eight-layer with four different directions is applied on the cylindrical surface. In addition, four-layers on the end-cap surface are obtained. Then, the workpiece is cured in an oven with controlled pressure and temperature to generate the final product. Fig. 6. shows the final product after the curing process.

5. Conclusion

Composites have become an important component for aerospace sector due to its superior properties. Robotic fibre placement technique is extensively used to fabricate composite structures with high accuracy. The current work proposed an effective technique for fabricating a pressure vessel. This technique is based on an 8-DoF workcell. A 6-DoF robot is combined with two different mandrel tools to fabricate the pressure vessel surfaces. The robot end-effector trajectories are calculated based on a feedback control system. The workpiece is then cured to get the final product. The proposed technique has the potential to fabricate the oxidizer tank which can be applied in different applications, e.g. (Hybrid rocked). In future work, the two parts of the pressure vessel will be fabricated using the robot operated with only one mandrel tool. A 2-DoF mandrel tool will be built to lay-up the composite material on the whole structure.

IOP Conf. Series: Materials Science and Engineering **859** (2020) 012018 doi:10.1088/1757-899X/859/1/012018

References

[1] S. Taj, M. A. Munawar, and S. Khan, "Natural fiber-reinforced polymer composites," *Proceedings-Pakistan Academy of Sciences,* vol. 44, no. 2, pp. 129, 2007.
[2] R. M. Jones, *Mechanics of composite materials*: CRC press, 2014.
[3] P. Zhao, B. Shirinzadeh, Y. Shi, S. Cheuk, and L. Clark, "Improved uniform degree of multi-layer interlaminar bonding strength for composite laminate," *Journal of Reinforced Plastics and Composites,* vol. 36, no. 17, pp. 1211-1224, 2017.
[4] G. Mittal, K. Y. Rhee, V. Mišković-Stanković, and D. Hui, "Reinforcements in multi-scale polymer composites: Processing, properties, and applications," *Composites Part B: Engineering,* vol. 138, pp. 122-139, 2018.
[5] C. Soutis, "Carbon fiber reinforced plastics in aircraft construction," *Materials Science and Engineering: A,* vol. 412, no. 1-2, pp. 171-176, 2005.
[6] L. Pehlivan, and C. Baykasoğlu, "An experimental study on the compressive response of CFRP honeycombs with various cell configurations," *Composites Part B: Engineering,* vol. 162, pp. 653-661, 2019.
[7] C. Grant, "Automated processes for composite aircraft structure," *Industrial Robot: An International Journal,* vol. 33, no. 2, pp. 117-121, 2006.
[8] Z. Gurdal, B. Tatting, and K. Wu, "Tow-placement technology and fabrication issues for laminated composite structures." p. 2017.
[9] B. Shirinzadeh, G. Alici, C. W. Foong, and G. Cassidy, "Fabrication process of open surfaces by robotic fibre placement," *Robotics and Computer-Integrated Manufacturing,* vol. 20, no. 1, pp. 17-28, 2004.
[10] F. Bullock, S. Kowalski, and R. Young, "Automated prepreg tow placement for composite structures," *Advanced materials: The challenge for the next decade.,* vol. 1, pp. 734-745, 1990.
[11] M. N. Grimshaw, C. G. Grant, and J. M. L. Diaz, "Advanced technology tape laying for affordable manufacturing of large composite structures." pp. 2484-2494.
[12] T. Aized, and B. Shirinzadeh, "Robotic fiber placement process analysis and optimization using response surface method," *The International Journal of Advanced Manufacturing Technology,* vol. 55, no. 1-4, pp. 393-404, 2010.
[13] L. Yan, Z. C. Chen, Y. Shi, and R. Mo, "An accurate approach to roller path generation for robotic fibre placement of free-form surface composites," *Robotics and Computer-Integrated Manufacturing,* vol. 30, no. 3, pp. 277-286, 2014.
[14] G. Alici, B. Shirinzadeh, A. McConville, C. W. Foong, and M. Ang, "A mathematical model for a pneumatically actuated robotic fibre placement system," *Robotica,* vol. 20, no. 5, pp. 545-551, 2002.
[15] P. Zhao, B. Shirinzadeh, Y. Shi, S. Cheuk, and L. Clark, "Multi-pass layup process for thermoplastic composites using robotic fiber placement," *Robotics and Computer-Integrated Manufacturing,* vol. 49, pp. 277-284, 2018.
[16] T. Sofi, S. Neunkirchen, and R. Schledjewski, "Path calculation, technology and opportunities in dry fiber winding: a review," *Advanced Manufacturing: Polymer & Composites Science,* vol. 4, no. 3, pp. 57-72, 2018.
[17] G. Rousseau, R. Wehbe, J. Halbritter, and R. Harik, "Automated Fiber Placement Path Planning: A state-of-the-art review," *Computer-Aided Design and Applications,* vol. 16, no. 2, pp. 172-203, 2018.
[18] B. Shirinzadeh, C. Wei Foong, and B. Hui Tan, "Robotic fibre placement process planning and control," *Assembly Automation,* vol. 20, no. 4, pp. 313-320, 2000.
[19] B. Shirinzadeh, G. Cassidy, D. Oetomo, G. Alici, and M. H. Ang Jr, "Trajectory generation for open-contoured structures in robotic fibre placement," *Robotics and Computer-Integrated Manufacturing,* vol. 23, no. 4, pp. 380-394, 2007.

[20] C. Hély, L. Birglen, and W.-F. Xie, “Feasibility study of robotic fibre placement on intersecting multi-axial revolution surfaces,” *Robotics and Computer-Integrated Manufacturing,* vol. 48, pp. 73-79, 2017.
[21] M. Hassan, D. Liu, and D. Xu, “A Two-Stage Approach to Collaborative Fiber Placement through Coordination of Multiple Autonomous Industrial Robots,” *Journal of Intelligent \& Robotic Systems,* vol. 95, no. 3-4, pp. 915-933, 2018.
[22] Y. Zhao, Z. Han, Y. Zhang, and H. Fu, "Path Planning and Post-processing Algorithm for Fiber Placement of the Aircraft Tail." pp. 96-99.

Automatic Digital Fringe Projection for Advanced Micro-Scale Connector Manufacturing System

Ching-Hua Lu, Cheng-Yang Liu

Department of Biomedical Engineering, National Yang-Ming University, Taipei City, Taiwan

cyliu66@ym.edu.tw

Abstract. The digital fringe projection microscope has been investigated and used to estimate three-dimensional shape of micro-scale connector for advanced manufacturing system. This measurement system consists of digital projector, optical lenses, stereo microscope, and digital camera. In the software, the imaging program based on including black and white structure light, seven-step phase-shifting calculation, and path-independent phase unwrapping is well developed. The imaging results indicate that the three-dimensional shape of micro-scale connector is acquired by using this measurement system. The optical resolution of this measurement system is 3 μm and the measurement speed is 0.6 s. This measurement system has excellent performance including easy operation, fast measurement speed and high resolution. This measurement system can be applied to real-time three-dimensional shape detection in product processing of micro-scale connector.

1. Introduction

In recent years, three-dimensional (3-D) optical shape measurement has become a popular research subject [1-10]. As a significant contactless shape measurement technology, phase-shifting projected fringe profilometry is widely used in profile inspection, reverse engineering, and advanced manufacturing system [11-20]. Compared to other 3-D measurement methods, such as laser scanning, phase-shifting fringe projection has the advantage of real-time acquiring dense point clouds. 3-D surface calculation is typically based on triangulation known from photogrammetry. Finding homologous points from captured images by phase information is a critical work for fringe projection system. Current method for establishing is one camera and one projector. In order to simplify match corresponding points, epipolar geometry is introduced into fringe projection calculation. This method can be utilized to calibrate the extrinsic parameters and further calculate the points of 3-D surface. A potential advantage of fringe projection calculation is that the desired 3-D coordinates can be easily received by defining the resolution of phase raster. The phase correlation is used to calculate complex shape by combining fringe projection and photogrammetry. If all phase maps from camera are well captured, phase correlation can perform very well in finding homologous surface. However, in real applications, phase maps may inevitably represent abruption and discontinuous because of the complexity of object surface and the directions of camera. To solve the problem of finding homologous surface from the complex phase maps, we propose the optical lenses based on a stereo microscope. By converting fringe projection from optical lenses and stereo-microscope, homologous surface of micro-scale connector can be easily and exactly calculated by phase values.

Content from this work may be used under the terms of the Creative Commons Attribution 3.0 licence. Any further distribution of this work must maintain attribution to the author(s) and the title of the work, journal citation and DOI.
Published under licence by IOP Publishing Ltd

IOP Conf. Series: Materials Science and Engineering **859** (2020) 012019 doi:10.1088/1757-899X/859/1/012019

2. Principle of the system

For three-dimensional (3-D) measurement, a structured light is used by encoding seven fringe patterns and capturing the deformed patterns using a digital camera. Figure 1 shows the automatic digital fringe projection for advanced manufacturing system. The components of this system include a projector, a stereo-microscope, a set of optical lenses, a digital camera with 1920■1080 resolution and an industrial computer. A set of optical lenses includes a plano-convex lens, a biconcave lens, and a biconvex lens. The ray tracing calculation is used to obtain the position relationship of a set of optical lenses. The fast pattern switching is reached by the projection mechanism of the digital light processing. Seven fringe patterns are projected sequentially which permits to capture seven deformed patterns separately.

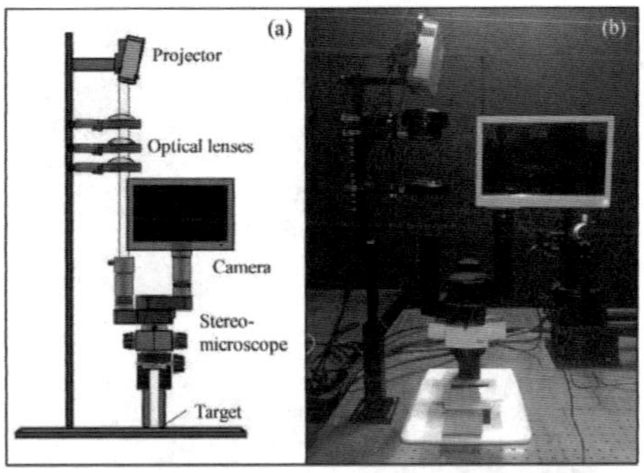

Figure 1. Automatic digital fringe projection for advanced manufacturing system: (a) schematic diagram and (b) picture.

Because the maximum numbers of fringe patterns are needed for calculations, the seven-step phase-shifting calculation is better for high-resolution measurement. A fringe pattern for different intensity distributions is defined by I_N. The intensity distribution of the Nth pattern is I_N. The average intensity of the background description is I'. The distributed amplitude of the pattern intensity is I''. The phase shift between each fringe pattern is $\pi/2$. The equations of seven fringe patterns are described as:

$$\begin{aligned} I_1 &= I' + I''\cos[\phi - \frac{3\pi}{2}] \\ I_2 &= I' + I''\cos[\phi - \pi] \\ I_3 &= I' + I''\cos[\phi - \frac{\pi}{2}] \\ I_4 &= I' + I''\cos[\phi] \\ I_5 &= I' + I''\cos[\phi + \frac{\pi}{2}] \\ I_6 &= I' + I''\cos[\phi + \pi] \\ I_7 &= I' + I''\cos[\phi + \frac{3\pi}{2}] \end{aligned} \tag{1}$$

The intensity of light source and fringe patterns is supposed in the same value. The phase in these fringe patterns can be retrieved by the equation (2).

$$\phi = \tan^{-1}[\frac{3I_3 + I_7 - I_1 - 3I_5}{4I_4 - 2I_2 - 2I_6}] \tag{2}$$

IOP Conf. Series: Materials Science and Engineering **859** (2020) 012019 doi:10.1088/1757-899X/859/1/012019

The phase value is a range of -π to +π in Eqs. (2) and (4). The continuous phase information is obtained by phase unwrapping calculation. We use quality guided path algorithm for unwrapping calculation. The quality guided path calculation is suitable for unwrapping from complex surfaces.

3. Results and discussion

A structured light is used by encoding seven fringe patterns for high-speed surface measurement. Figure 2 shows the seven-step fringe patterns projected on a micro-scale connector. Seven phase-shifted fringe patterns are used to reconstruct 3-D profile through phase wrapping, phase unwrapping and phase to dimension conversion processes. The wrapped phase map is obtained by applying seven-step phase-shifting algorithm.

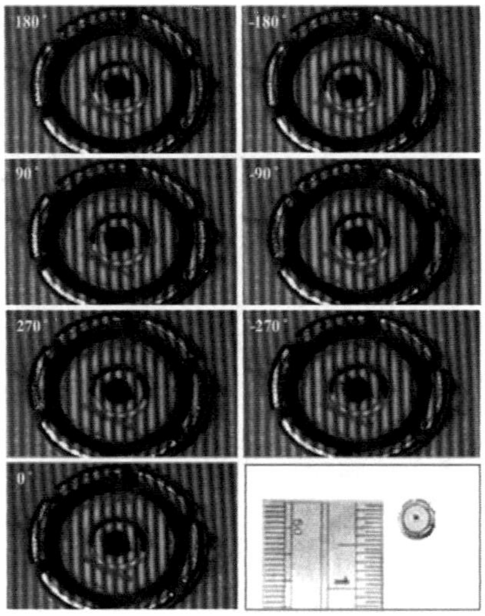

Figure 2. Seven-step fringe patterns projected on the surface of micro-scale connector. The insert in the bottom right is the picture of micro-scale connector.

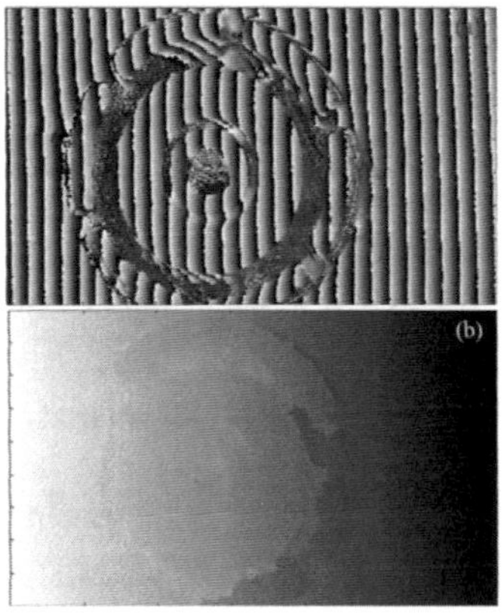

Figure 3. Seven-step phase-shifting calculations for micro-scale connector: (a) wrapped phase map and (b) unwrapped phase map.

Figure 3 shows the calculated wrapped phase map and unwrapped phase map by seven-step phase-shifting algorithm for micro-scale connector. This phase map indicates that the shape change strongly contributes to the wrapped phase information. Since the phase information in Fig. 3(a) is a periodic function of 2π, the unwrapped phase map is obtained by phase unwrapping calculation. Figure 3(b) illustrates the unwrapped phase map by quality guided path calculation for micro-scale connector. The absolute phase map is a continuous value in which the highest value shows a white point and the

IOP Conf. Series: Materials Science and Engineering **859** (2020) 012019 doi:10.1088/1757-899X/859/1/012019

lowest value shows a black point. The absolute phase map is converted into real 3-D shape by using the calibrated coefficients.

The distorted fringe patterns are used to reconstruct 3-D shape through phase wrapping, phase unwrapping and phase to real size conversion. Using the proposed digital fringe projection system, micro-scale connector is captured and processed to obtain the 3-D representation. Figure 4 shows the 3-D representation of the captured micro-scale connector. The experimental results clearly show that the 3-D imaging system based on fringe projection technique correctly obtains the 3D surface of micro-scale connector. The seven-step phase-shifting calculation obviously mended the smoothness of the connector surface. The nonlinear phase errors in the 3-D representation are balanced by quality guided path unwrapping. The measurement capabilities of complexity and high-resolution for our digital fringe projection system are well proven.

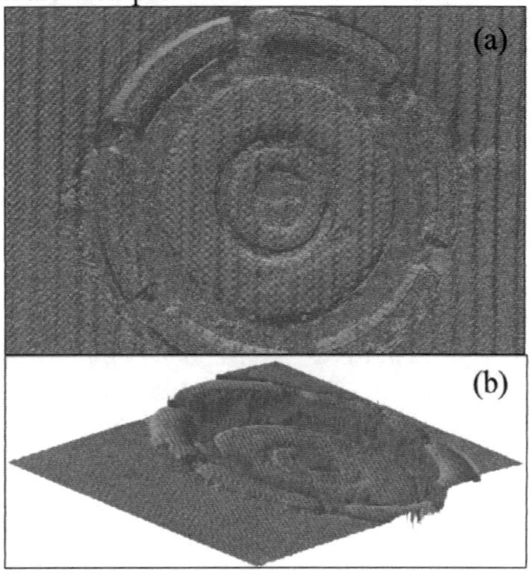

Figure 4. 3-D representation of micro-scale connector.

4. Conclusion

In this research, a high-speed digital fringe projection microscope is proposed to measure 3-D surface of micro-scale connector. The optical source used in this microscope is structured light with fringe patterns. A projector is used as optical source to project structured light onto the connector surface. The distorted fringe patterns are imaged by the digital camera. The seven-step phase-shifting and quality guided path unwrapping calculations are used to compute the phase values. The digital camera is controlled by using an industrial computer. A complete 3-D surface of micro-scale connector is obtained by our digital microscope. The phase acquisition, shape reconstruction and exhibition have performed simultaneously at a speed of 0.6 s. The experimental results may provide a high-resolution and real-time 3-D shape detection for micro-scale connector in advanced manufacturing system.

Acknowledgments

The authors would like to appreciate the partial financial support from the Ministry of Science and Technology in Taiwan with the project number of MOST 108-2221-E-010-012-MY3.

References

[1] P. S. Huang, Q. J. Hu, and F. P. Chiang, "Double three-step phase-shifting algorithm," Appl. Opt., Vol. 41, No. 22, pp. 4503-4509, 2002.

[2] S. Fang, L. Wang, M. Komori, A. Kubo, "Simulation method for interference fringe patterns in measuring gear tooth flanks by laser interferometry," Appl. Opt., Vol. 49, pp. 6409-6415, 2010.

[3] J. Li, O. Sasaki, "Measurement of a cross-sectional profile of a thread gauge using a sinusoidally vibrating interference pattern," Appl. Opt., Vol. 50, pp. 3470-3474, 2011.

[4] Z. Zhang, S. Huang, Y. Xu, C Chen, Y. Zhao, N. Gao, and Y. Xiao, "3D palmprint and hand imaging system based on full-field composite color sinusoidal fringe projection technique," Appl. Opt., Vol. 52, No. 25, pp. 6138-6145, 2013.
[5] C. Zuo, Q. Chen, G. Gu, S. Feng, F. Feng, R. Li, and G. Shen, "High-speed three-dimensional shape measurement for dynamic scenes using bi-frequency tripolar pulse-width-modulation fringe projection," Opt. Lasers Eng., Vol. 51, pp. 953-960, 2013.
[6] S. Huang, Z. Zhang, Y. Zhao, J. Dai, C. Chen, Y. Xu, E. Zhang, and L. Xie, "3D fingerprint imaging system based on full-field fringe projection profilometry," Opt. Lasers Eng., Vol. 52, pp. 123-130, 2014.
[7] S. Feng, Y. Zhang, Q. Chen, C. Zuo, R. Li, and G. Shen, "General solution for high dynamic range three-dimensional shape measurement using the fringe projection technique," Opt. Lasers Eng., Vol. 59, pp. 56-71, 2014.
[8] H. Nguyen, D. Nguyen, Z. Wang, H. Kieu, and M. Le, "Real-time, high-accuracy 3D imaging and shape measurement," Appl. Opt., Vol. 54, pp. A9-A17, 2015.
[9] H. Wang, Q. Kemao, and S. Soon, "Valid point detection in fringe projection profilometry," Opt. Express, Vol. 23, pp. 7535-7549, 2015.
[10] Z. Lei, C. Wang, and C. Zhou, "Multi-frequency inverse-phase fringe projection profilometry for nonlinear phase error compensation," Opt. Lasers Eng., Vol. 66, pp. 249-257, 2015.
[11] C. Liu, and T. Yen, "Digital multi-step phase-shifting profilometry for three-dimensional ballscrew surface imaging," Opt. Laser Technol., Vol. 79, pp. 115-123, 2016.
[12] M. Wang, Y. Yin, D. Deng, X. Meng, X. Liu and X. Peng, "Improved performance of multi-view fringe projection 3D microscopy," Opt. Express, Vol. 25, pp. 19408-19421, 2017.
[13] M. Dai, F. Yang, C. Liu, and X. He, "A dual-frequency fringe projection three-dimensional shape measurement system using a DLP 3D projector," Opt. Commun., Vol. 382, pp. 294-301, 2017.
[14] G. Parra Escamill, F. Kobayashi, and Y. Otani, "Three-dimensional surface measurement based on the projected defocused pattern technique using imaging fiber optics," Opt. Commun., Vol. 390, pp. 57-60, 2017.
[15] C. Chen, N. Gao, X. Wang, and Z. Zhang, "Adaptive projection intensity adjustment for avoiding saturation in three-dimensional shape measurement," Opt. Commun., Vol. 410, pp. 694-702, 2018.
[16] S. Yu, J. Zhang, X. Yu, X. Sun, H. Wu, and X. Liu, "3D measurement using combined Gray code and dual-frequency phase-shifting approach," Opt. Commun., Vol. 413, pp. 283-290, 2018.
[17] Z. Wang, Z. Zhang, N. Gao, Y. Xiao, F. Gao, and X. Jiang, "Single-shot 3D shape measurement of discontinuous objects based on a coaxial fringe projection system," Appl. Opt., Vol. 58, No. 5, pp. A169-A178, 2019.
[18] S. Gai, F. Da, and M. Tang, "A flexible multi-view calibration and 3D measurement method based on digital fringe projection," Meas. Sci. Technol., Vol. 30, pp. 025203, 2019.
[19] Y. Ma, D. Yin, C. Wei, S. Feng, J. Ma, S. Nie, and C. Yuan,, "Real-time 3-D shape measurement based on radial spatial carrier phase shifting from circular fringe pattern," Opt. Commun., Vol. 450, pp. 6-13, 2019.
[20] Y. Xu, H. Zhao, H. Jiang, and X. Li, "High-accuracy 3D shape measurement of translucent objects by fringe projection profilometry," Opt. Express, Vol. 27, pp. 18421-18434, 2019.

Application Phases for Productivity Improvement through Lean Methods Assessments in an Aeronautical Company – Case Study

Leandro Silvério, Luís Gonzaga Trabasso, Marcus Vinicius Pereira Pessôa

Aeronautics Institute of Technology (ITA), Praça Marechal Eduardo Gomes, 50—Vila das Acácias, 12228–900 São José dos Campos/SP – Brazil

leandro.silverio.pereira@gmail.com

Abstract. A well defined manufacturing company focuses on lean practices along the whole product development process, not exclusively during the manufacturing and production phases. This study focuses on the application steps to be adopted for productivity improvement trough lean methods assessments in a product development company for the aeronautical industry. It also aims to provide insights for decision makers along the whole lean transformation patch, and to reduce the waste generation by eliminating the non–value–added activities resulted by a bad execution of the lean implementation tools and methods. By deploying the proposed study, the assessed enterprise increased its overall lean engagement level and production outputs.

1. Introduction

The concept of lean manufacturing (LM) was popularized by Womack et al. [1] by extracting the core lean elements from the automotive production benchmarking. LM is regarded as a successor of the Toyota Production System (TPS), a business philosophy that has been in development for over 40 years. Lean production (LP) stems from the Japanese TPS [2] during the hard times of the post–war [3]; that Taiichi Ohno and Eiji Toyoda both Japanese engineers, developed between 1948 and 1975. The essence of TPS is a mindset that embraces continuous improvement (CI) and the capacity to adopt a set of values and culture, as summarized in the TPS house diagram by Liker [4]. The house metaphor is illustrated in the Figure 1 and demonstrates that a house is strong only if the roof, pillars and foundation are also strong. As far as developments may suffer from failures, delays and/or cost overruns [5], the TPS house shows that having a true LM requires adaptations in several sectors [6], moreover it is composed by multiple economic and operational performance dimensions [7] and represents a manufacturing culture of CI and waste reduction through participation of all employees [8], that must abide by a standardized processes [9] in an environment where their attitudes can reflect reactions to change [10]. Nowadays, competitions in manufacturing firms continue to grow [11]. Thus, from the recent years LM has been widely tested in recent literature reviews [12] and practitioners throughout the world by increasing its adoption for the industry 4.0 [13-14]; the implementation of LM emerging from the supply chain processes [15] and management [16-17]; and also by evaluating its motivational influence on the lean way [18-21]. In this paper, the focus is given to the application phases for productivity improvement through lean methods assessments in an aeronautical company. Thus, the case study enterprise was assessed as a complex socio–technical lean system.

Content from this work may be used under the terms of the Creative Commons Attribution 3.0 licence. Any further distribution of this work must maintain attribution to the author(s) and the title of the work, journal citation and DOI.
Published under licence by IOP Publishing Ltd

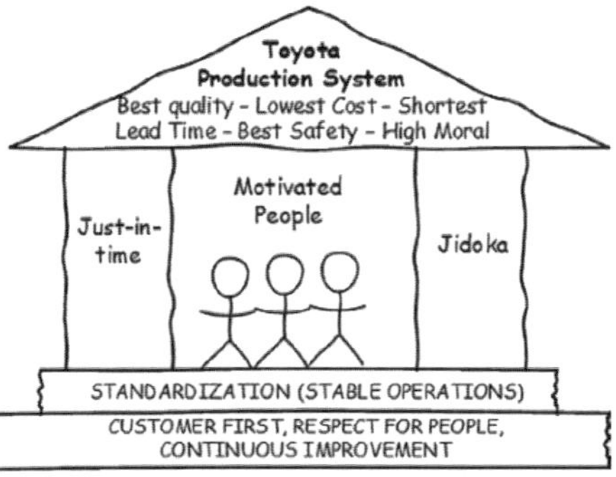

Figure 1. The TPS house. [6]

2. Background

As product complexity increases, the trade–offs become increasingly complex [22]. Worldwide air traffic has been increasing through the last years, and recent researches indicate that the demand for new aircraft will continue to rise over next few decades [23]. Moreover, the manufacturing industry each time moves towards to products customization [24], while the time–to–market dimension has shrunk [25] in a scenario where aircraft design is largely dependent on previous design experience [26] and different lean strategies are utilized by manufacturing industries to improve its performance [27]. Brazil as an emerging country needs to catch up with technology to extend its international position, especially in the space sector [28]. Thus, the selected case study enterprise emerges from this context, and acts in the aeronautical industry for the aeronautical components supply industry in Brazil. The company was founded on July 4, 1990 at Rio de Janeiro and had around 100 employees by the time the method was deployed. The facility contained manufacturing, laboratory and engineering engaged in the development and production of aeronautical hoses and hydro–mechanical components.

LM plays a critical role in reducing waste when combined with other production resources [29] and do can bring an overall change in a manufacturing plant [30]. To deeply understand the manufacturing process is vital for tracking the bottlenecks and then, the area subject to improvements [31]. In other words, as far as work standardization is part of LM [32] and to save the company's resources, it is necessary to identify the right set of metrics [33] prior to the method application; thus a plant tour was performed at the facility in order to visually check its ongoing lean processes. The top management and leadership commitment is crucial for the lean adoption [34]; that way the plant tour was followed by all personnel in charge of the lean implementation. A few examples of existing lean approaches resulted from the plant tour in the case study enterprise are shown in the Figure 2 and Figure 3 below:

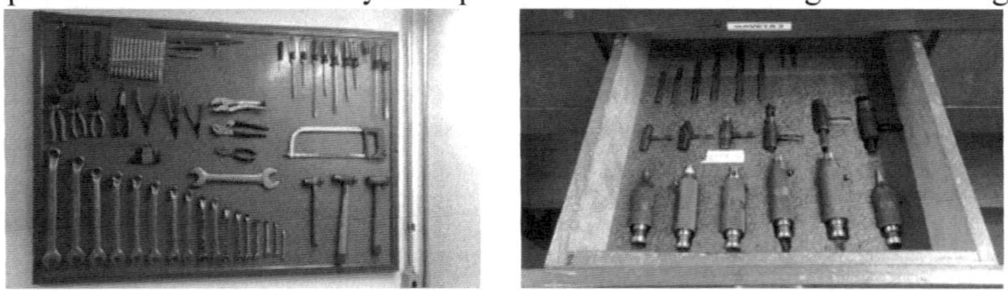

Figure 2. 5S manufacturing tools.

Figure 3. Kanban manufacturing chart and stock control placard.

IOP Conf. Series: Materials Science and Engineering **859** (2020) 012020 doi:10.1088/1757-899X/859/1/012020

3. Methodology

The application phases for productivity improvement through lean methods assessments in an aeronautical company are indicated in the Figure 4. The selected lean assessment method for the case study enterprise is the framework proposed by Silvério et al. [35], which includes all mathematics, calculations and roadmap definition for the decision-making process criteria. The activities coordination along the interaction cycle execution was sponsored by the company's top management along the lean way (LW), including a key–person named "facilitator" or "lean agent," which is the collaborator responsible to guarantee all adequate resources along the lean implementation patch.

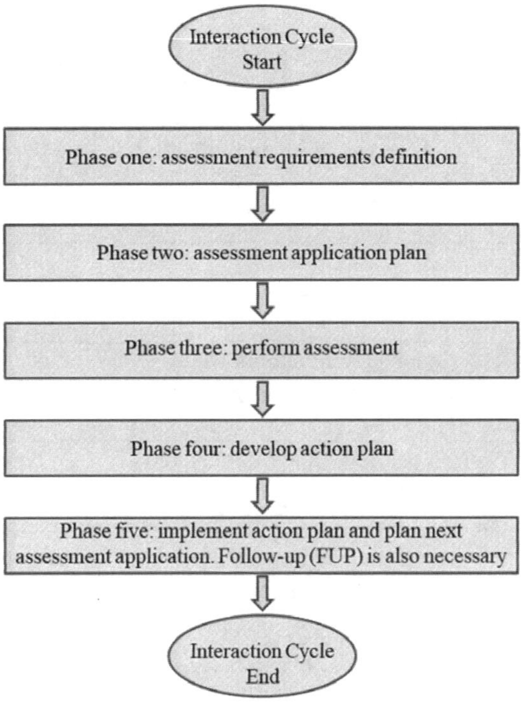

Figure 4. Lean journey method application phases interaction cycle (constructed by authors).

Each phase from the flowchart presented in the Figure 4 has its own pre–defined activities breakdown; later on detailed in the Table 1. For each activity breakdown, a task owner is in charge of its deployment and coordination in all organizational levels. Some of those refereed activities can also have a "multi–task team" defined, by assuming one or more roles according to the following criteria: leader, facilitator (lean agent) or employee. The activities breakdown for each phase is defined below.

Phase one. Phase one starts the interaction cycle for the application phases. In the phase one, it is expected to obtain leadership commitment; to define the method implementation schedule; to define the participant's roles and responsibilities; to allocate resources for the method implementation and application and; to review the progress implementation of current action plans (if applicable). Phase one is considered a key–phase for the productivity improvement through lean methods application and shall have a high involvement from the leadership and the facilitator (lean agent) along its execution.

Phase two. In the phase two, it is expected to identify the respondents; and to perform the method introduction and training. Phase two is also considered a strategic–phase for the productivity improvement through lean methods application and shall have a close participation and engagement from all organizational levels directly involved in the lean transformation patch.

Phase three. In the phase three, it is expected to conduct the assessment; to collect and process the results; and to discuss, analyze and evaluate the results. Phase three is considered an operational phase, where the strategy defined in the previous phases take place along its deployment and application.

Phase four. In the phase four, it is expected to develop the action plan and to prioritize the resources. In the phase four, some results can already emerge and thus, it shall also have a high involvement from the leadership and the facilitator (lean agent) along its execution.

Phase five. In the phase five, it is expected to implement the defined action plan, record the results and plan the next assessment application. Phase five closes one interaction cycle for the application phases for productivity improvement through lean methods in an aeronautical company. As per the previous phases, it is also expected a close participation from all organizational levels directly involved in the lean transformation patch along its execution.

Table 1 presents the method application phases interaction cycle responsibilities distribution, where the lean assessment framework from Silvério et al. [35] was deployed and assisted by the leadership for the case study enterprise. As a result, the leanness level was defined, associating its results to a lean roadmap as per the chosen method's criteria, to be used as guidance for managers to introduce recommended changes on their lean implementation patch, and also to prepare for the next interaction cycle application as per the Figure 4 lean journey method application phases interaction cycle.

Table 1. Lean journey method application phases interaction cycle responsibilities distribution.

Phases	Tasks	Personnel in Charge		
		Leader	**Facilitator (Lean Agent)**	**Employee**
Phase one: assessment requirements definition	To obtain leadership commitment	X	X	
	To define the method implementation schedule	X	X	
	To define the participant's roles and responsibilities		X	
	To allocate resources for the method implementation and application	X		
	To review the progress implementation of current action plans (if applicable)	X	X	
Phase two: assessment application plan	To identify the respondents		X	X
	To perform the method introduction and training	X	X	X
Phase three: perform assessment	To conduct the assessment		X	
	To collect and process the results		X	
	To discuss, analyze and evaluate the results	X	X	
Phase four: develop action plan	To develop the action plan and to prioritize the resources	X	X	X
Phase five: implement action plan and plan next assessment application. Follow–up is also necessary.	To implement the defined action plan, record the results and plan the next assessment application	X	X	X

4. Results and analysis

Adopting lean practices are not easy, particularly, in small and medium–sized enterprises (SMEs) [36-37]; that is also the criteria fulfilled by the case study firm. Moreover, lean solutions bring benefits for companies [38-40] and consequently, to society. That way, during three months after deploying the chosen lean assessment method application in the case study enterprise, the current lean engagement rate was observed and recorded over three full interaction cycles execution as per the Figure 4 flowchart. The trend results observed after each interaction cycles are summarized in the Table 2.

Table 2. Interaction cycles summary for the case study enterprise.

Interaction Cycle 1*	Interaction Cycle 2*	Interaction Cycle 3*
50% (regular)	60% (regular)	70% (good)

* Current lean engagement level (measured rate) compared to an ideal target scenario (100%).

As demonstrated in the Table 2, the case study enterprise improved its overall lean index from 50% (regular) measured in the first interaction cycle to 70% (good) recorded in the last one by deploying the framework proposed by Silvério et al. [35] as per the method application phases interaction cycle presented in the Figure 4, divided according to the phases responsibilities distribution presented in the Table 1. The elements "sponsored" by leadership had a bigger impact on the lean assessment method application results, demonstrating that lean initiatives are closely related to all enterprise's levels of engagement to succeed on the lean implementation patch. Improvements on the production line were also observed along the interaction cycle's execution, increasing its production output as far as the chosen framework for the case study application also considers elements such as: process, tools and technology, waste and CI as "key lean elements" on its development and construction.

As an overall result, the application phases for productivity improvement through lean methods assessments in an aeronautical company proposed by this study has supported the efficiency and productivity increase for the case study enterprise on the lean way perspective. Although the resulted efficiency and productivity increment may seem quite small, the refereed overall impact had improved the company's production output in a sustainable manner, as far as the case study enterprise continued to perform the interactions cycles along its lean journey.

5. Conclusions

Even though having a considerable number of information being delivered by the application phases' deployment in a real aeronautical company, it is not possible to have a failure–proof lean company for every single process. Lean basis depends on people; and people are not a failure–proof resource. Thus, a failure–proof lean organization is not a feasible, nor a realistic scenario to be considered. In this study, a real problem application demonstrated productivity improvements through lean methods assessments application in an aeronautical company. As a result, the interaction cycle and responsibilities distribution demonstrated to be a useful support tool for decision makers to eliminate the non–value–added activities during the deployment phases of lean implementation practices, methods and processes in companies emerging as true lean organizations.

By applying the proposed method the assessed enterprise presented its real phases and responsibilities distribution for the interaction cycles execution as demonstrated in the Table 1. That way, the acquired data can also serve as reference information for future studies to be developed on the lean field, as far as the same responsibilities and tasks distribution can also be replicated for different industrial segments. Moreover, the method deployment in different industrial sectors, besides the aeronautical one, seems to be a particularly interesting area for further researches in order to perform a benchmarking analysis and comparative approaches on similar tools, methods and processes. A digital platform for each phase application could also be developed in order to provide a real–time key performance indicator (KPI), to be used by decision makers and managers in order to facilitate the data-mining and data-sharing information, among all personnel involved on the company's lean transformation patch.

Acknowledgments
The author(s) declared no potential conflicts of interest with respect to the research, authorship, and/or publication of this article. The author(s) received no financial support for the research, authorship, and/or publication of this article.

References

[1] Womack JP, Jones DT and Roos D (1991) *The Machine that Changed the World.* New York: Harper Perennial.

[2] Chiarini A, Baccarani C and Mascherpa V (2018) Lean production, Toyota Production System and Kaizen philosophy: A conceptual analysis from the perspective of Zen Buddhism. *The TQM Journal* 30. DOI: doi.org/10.1108/TQM-12-2017-0178.

[3] Shingo S (1981) *Study of the Toyota production system from an industrial engineering viewpoint.* New York: Productivity Press. DOI: doi.org/10.4324/9781315136509.

[4] Liker JK (2004) *The Toyotaway 14 management principles from Worlds Greatest Manufactures.* New York: McGraw–hill.

[5] Wekerle T, Trabasso LG, Loures L, Villela T, Brandão A and Leonardi R (2017) Design for Autonomy: Integrating Technology Transfer into Product Development Process. *Journal of Industrial Integration and Management* 02. 1750004. DOI: doi.org/10.1142/S242486221750004X.

[6] Pessôa MVP and Trabasso LG (2017) *The Lean Product Design and Development Journey: A Practical View.* Berlin: Springer. DOI: doi.org/10.1007/978-3-319-46792-4.

[7] Shah R and Ward PT (2003) Lean Manufacturing: Context, Practice Bundles, and Performance. *Journal of Operations Management* 21 129–149. DOI: doi.org/10.1016/S0272-6963(02)00108-0.

[8] Suhardi B, Anisa N and Laksono PW (2019) Minimizing waste using lean manufacturing and ECRS principle in Indonesian furniture industry. *Cogent Engineering* 6. DOI: doi.org/10.1080/23311916.2019.1567019.

[9] Alkhoraif AA, McLaughlin P and Rashid H (2019) A framework to improve lean implementation by review leveraging aspects of organisational culture: the case of Saudi Arabia. *International Journal of Agile Systems and Management* 12 124. DOI: doi.org/ 10.1504/IJASM.2019.10018747.

[10] Mcmackin J and Flood PC (2019) A theoretical framework for the social pillar of lean. *Journal of Organizational Effectiveness: People and Performance* 6. DOI: doi.org/10.1108/JOEPP-06-2018-0039.

[11] Susanto TA and Nur R (2017) Experimental research to investigate the performance of bio coolant when turning of mild carbon steel. In: *International Conference on Manufacturing, Material and Metallurgical Engineering (ICMMME).* Bangkok, Thailand, 17–19 March 2017. IOP Publishing. DOI: doi.org/10.1088/1757-899X/191/1/012028.

[12] Danese P, Manfè V and Romano P (2017) A Systematic Literature Review on Recent Lean Research: State-of-the-art and Future Directions. *International Journal of Management Reviews.* DOI: doi.org/10.1111/ijmr.12156.

[13] Rossini M, Costa F, Tortorella GL et al. (2019) The interrelation between Industry 4.0 and lean production: an empirical study on European manufacturers. *International Journal of Advanced Manufacturing Technologies* 102 3963–3976. DOI: doi.org/10.1007/s00170-019-03441-7.

[14] Romero D, Gaiardelli P, Powell D, Wuest T and Thurer M (2019) Rethinking Jidoka Systems under Automation & Learning Perspectives in the Digital Lean Manufacturing World. In: *9th IFAC Conference on Manufacturing Modelling, Management and Control.* Berlin, Germany, 28–30 August 2019. IFAC PapersOnLine Series. DOI: doi.org/10.1016/j.ifacol.2019.11.309.

[15] Meng X (2019) Lean management in the context of construction supply chains. *International Journal of Production Research* 57 1–15. DOI: doi.org/10.1080/00207543.2019.1566659.

[16] Sony M (2019) Lean Supply Chain Management and Sustainability. *Advances in Logistics, Operations, and Management Science.* DOI: doi.org/10.4018/978-1-5225-8970-9.ch004.

[17] Tortorella GL, Miorando RF and Marodin GA (2017) Lean Supply Chain Management: Empirical research on practices, contexts and performance. *International Journal of Production Economics* 193. DOI: doi.org/10.1016/j.ijpe.2017.07.006.

[18] Castro F, Figueiredo SP, Pereira–Guizzo C and Passos FU (2019) Effect of the motivational factor on lean manufacturing performance: the case of a multinational consumer goods company. *Gestão & Produção* 26. DOI: doi.org/10.1590/0104-530x4850-19.
[19] Christensen R, Greenhalgh S and Thomassen Anja (2019) When a business case is not enough, Motivation to work with Lean. *Lean Construction Journal (LCJ)* 121–130. DOI: doi.org/10.24928/2019/0146.
[20] Schmidt G (2019) The need for goal–setting theory and motivation constructs in Lean management. *Industrial and Organizational Psychology* 12. DOI: doi.org/10.1017/iop.2019.48.
[21] Herakovic N, Metlikovič P and Debevec M (2014) Motivational Lean Game to Support Decision between Push and Pull Production Strategy. *International Journal of Simulation Modelling* 13. 443–446. DOI: doi.org/10.2507/IJSIMM13(4)4.275.
[22] Borgue O, Panarotto M and Ola I (2019) Modular product design for additive manufacturing of satellite components: Maximizing product value using genetic algorithms. *Concurrent Engineering: Research and Applications*. DOI: doi.org/10.1177/1063293X19883421.
[23] Silva A and Trabasso LG (2019) Design for Automation within the aeronautical domain. *Journal of the Brazilian Society of Mechanical Sciences and Engineering*. DOI: doi.org/10.1007/s40430-019-1791-y.
[24] Anumba CJ, Baldwin AN, Bouchlaghem D, Prasad B, Cutting–Decelle Af, Dufau J and Mommessin M (2000) Integrating Concurrent Engineering Concepts in a Steelwork Construction Project. *Concurrent Engineering: Research and Applications* 8(3): 199–212. DOI: doi.org/10.1177/1063293X0000800304.
[25] Prasad B (1997) Re–engineering life–cycle management of products to achieve global success in the changing marketplace. *Industrial Management & Data Systems* 97(3): 90–98. DOI: doi.org/10.1108/02635579710173158.
[26] Fansuri AFH, Rose ANM, Ab. Rashid MFF, Mohamed Nik NMZ and Ahmad H (2018) Productivity Improvement Through Line Balancing at Electronic Company – Case Study. In: *International Conference on Manufacturing, Material and Metallurgical Engineering (ICMMME)*. Kuala Lumpur, Malaysia, 17–19 March 2018. IOP Publishing. DOI: doi.org/10.1088/1757-899X/409/1/012015.
[27] Singh J, Singh H and Singh G (2018) Productivity Improvement using lean manufacturing in manufacturing Industry of Northern India – A Case Study. *International Journal of Productivity and Performance Management* 67. 00–00. DOI: doi.org/10.1108/IJPPM-02-2017-0037.
[28] Wekerle T, Trabasso LG and Loures L (2019) Technology Nationalization in the Space Sector: The Brazilian Perspective. In: *Systems Engineering in Research and Industrial Practice*. Springer–Cham (Switzerland). Stjepandić J., Wognum N., J. C. Verhagen W. (eds). DOI: doi.org/10.1007/978-3-030-33312-6_12.
[29] Bhasin S (2015) *Lean management beyond manufacturing: A holistic approach*. Springer–Cham (Switzerland), pp: 149–160. DOI: doi.org/10.1007/978-3-319-17410-5.
[30] Puvanasvaran AP, HuiHui Y and Norazlin N (2014) Managing waste elimination database in Lean manufacturing: Improve problem solving capability. *American Journal of Engineering and Applied Sciences* 7 (2): 271–281 (2014). DOI: doi.org/10.3844/ajeassp.2014.271.281.
[31] Badeeb AM, Abdulaal R and Bafail AO (2019) An Application of Lean Manufacturing Techniques in Paint Manufacturing Company: A Case Study. *Journal of King Abdulaziz University* 28(2):51-73. DOI: doi.org/10.4197/Eng.28-2.5.
[32] Liu F, Li Q and Geng H (2019) Correlation Analysis between Gross Parameters and Performance Parameters of the Light Aircraft. In: *International Conference on Manufacturing, Material and Metallurgical Engineering (ICMMME)*. Chengdu, China, 22–25 March. IOP Publishing. DOI: doi.org/10.1088/1757-899X/538/1/012050.
[33] Acosta LMC, Trabasso LG and de Junior CSA (2003) Analysis of the Balanced Scorecard Formulation Process for Setting up Engineering Performance Metrics. In: *14th International Conference on Engineering Design*. Stockholm, Sweden, August 19–21.

[34] Alnajem MN, Garza–Reyes JA and Antony J (2019) Lean readiness within emergency departments: a conceptual framework. *Benchmarking: An International Journal* 26. DOI: doi.org/10.1108/BIJ-10-2018-0337.

[35] Silvério L, Trabasso LG and Pessôa MVP (2019) A Roadmap for a Leanness Company to Emerge as a True Lean Organization. *Concurrent Engineering: Research and Applications*. 28(1), 3–19. DOI: doi.org/10.1177/1063293X19888259.

[36] Cowger G (2016) Half measures gets less than half results. *Mechanical Engineering* 138(1), 30–35. DOI: doi.org/10.1115/1.2016-Jan-1.

[37] Hu Q, Mason R, Williams S and Found P (2015) Lean implementation within SMEs: A literature review. *Journal of Manufacturing Technology Management* 26(7), 980–1012. DOI: doi.org/10.1108/JMTM-02-2014-0013.

[38] Alves AC, Dinis–Carvalho J, Sousa RM, Moreira F and Lima RM (2011) Benefits of lean management: Results from some case studies in Portugal. In: *Proceedings of 6º Congresso Luso–Moçambicano de Engenharia (CLME2011).* Maputo, Mozambique, South African, 29 August – 2 September 2011. INEGI.

[39] Alves AC, Kahlen FJ, Flumerfelt S and Siriban–Manalang AB (2014) Fostering sustainable development thinking through lean engineering education. In: *Proceedings of the ASME 2014 International Mechanical Engineering Congress and Exposition*. Montreal, Quebec, Canada, 14–20 November 2014. ASME. DOI: doi.org/10.1115/IMECE2014-38192.

[40] Resende V, Alves AC, Batista A and Silva Â (2014) Financial and human benefits of lean production in the plastic injection industry: An action research study. *International Journal of Industrial Engineering and Management* 5(2), 61–75.

AUTHOR INDEX